EFFICIENT MATERIALS
AND COATINGS APPLICATIONS
FOR IMPROVED DESIGN AND
CORROSION RESISTANCE

OTHER BOOKS IN THE ASM
MATERIALS/METALWORKING TECHNOLOGY SERIES:

Cutting Tool Materials
Residual Stress for Designers and Metallurgists
Process Modeling — Fundamentals and Applications to Metals
Aluminum Transformation Technology and Applications
Machinability Testing and Utilization of Machining Data
Applications of Lasers in Materials Processing
Coatings for Corrosion Prevention
Prevention of Structural Failures — The Role of NDT, Fracture Mechanics and Failure Analysis
Nondestructive Evaluation in the Nuclear Industry
Welding of HSLA (Microalloyed) Structural Steels
What Does the Charpy Test Really Tell Us?
Control of Distortion and Residual Stress in Weldments
Influence of Metallurgy on Machinability
Sulfide Inclusions in Steel
Influence of Metallurgy on Hole Making Operations

Library of Congress Cataloging in Publication Data

Main entry under title:

Efficient materials and coatings applications for
improved design and corrosion resistance.

(Materials/metalworking technology series)
Includes bibliographical references.
1. Motor vehicles--Materials--Congresses.
2. Motor vehicles--Corrosion--Congresses. I. Vande Walle, Larry J. II. Materials & Processing Congress (1979: Chicago, Ill.) III. American Society for Metals. Highway and Off-Highway Vehicles Activity. IV. Series.
TL154.E46 629.2'32 81-3440
ISBN 0-87170-107-3 AACR2

PRINTED IN THE UNITED STATES OF AMERICA

EFFICIENT MATERIALS AND COATINGS APPLICATIONS FOR IMPROVED DESIGN AND CORROSION RESISTANCE

Proceedings of the
American Society for Metals
Highway and Off-Highway Vehicles Activity Sessions
Materials & Processing Congress
13-15 November 1979
Chicago, Illinois

Sponsored by the
Highway and Off-Highway Vehicles Activity of the ASM
Materials Systems and Design Division

Organized and edited by
LARRY J. VANDE WALLE
International Harvester Company

MATERIALS/METALWORKING TECHNOLOGY SERIES

American Society for Metals
Metals Park, Ohio 44073

PREFACE

The Technical Divisions of ASM were responsible for programming the 1979 Materials and Processing Congress. In particular, the Highway and Off-Highway Vehicles Activity of the Materials Systems and Design Division organized the four sessions covered by these proceedings.

The sessions were planned to expand on topic matter addressed in previous years or, as in one case, address a different topic. Composites, rare earth treated steel, nodular iron castings and aluminum castings were all discussed as they have been in the past; however, a few new concepts were involved. There were similar presentations on paint protection, surface preparation and zinc containing coatings. The new topic matter addressed at this Congress related to liquid retaining and sealing materials. Subject matter ranged from sealing porosity to gasketing to retaining fasteners.

Thanks are expressed to all of the authors, session chairmen and their companies. Without their willing dedication, the stage would have been empty. All compliments on these sessions received by the Technical Divisions must be reflected on these people. Special thanks also to the committee members for helping to organize these sessions. They are: F. E. Arant, the Timken Company; D. L. Cheever, Deere & Company; J. P. Faunce, Martin Marietta Laboratories; D. W. Gomersall, Inland Steel Research Laboratories; B. A. Juel, Funk Manufacturing; J. E. Miller, Delco Products Division; W. R. Phillips, Investigative Engineering Laboratories; R. H. Van Pelt, Caterpillar Tractor Company; and W. B. Young, Dana Corporation.

Thanks also to Elliot Boardman and Patti Gaeta at ASM headquarters for their support and assistance.

Larry J. Vande Walle
International Harvester Company
Committee Chairman

CONTENTS

DESIGN EFFICIENCY — I: HIGHWAY AND OFF-HIGHWAY VEHICLES

DESIGN EFFICIENCY — II: ADHESIVES AND SEALANTS IN THE STRUCTURAL WORLD

* Paper not available.

* Paper not available.

Library of Congress Cataloging in Publication Data

Main entry under title:

Efficient materials and coatings applications for
improved design and corrosion resistance.

(Materials/metalworking technology series)
Includes bibliographical references.
1. Motor vehicles--Materials--Congresses.
2. Motor vehicles--Corrosion--Congresses. I. Vande Walle, Larry J. II. Materials & Processing Congress (1979: Chicago, Ill.) III. American Society for Metals. Highway and Off-Highway Vehicles Activity. IV. Series.
TL154.E46 629.2'32 81-3440
ISBN 0-87170-107-3 AACR2

PRINTED IN THE UNITED STATES OF AMERICA

EFFICIENT MATERIALS AND COATINGS APPLICATIONS FOR IMPROVED DESIGN AND CORROSION RESISTANCE

Proceedings of the
American Society for Metals
Highway and Off-Highway Vehicles Activity Sessions
Materials & Processing Congress
13-15 November 1979
Chicago, Illinois

Sponsored by the
Highway and Off-Highway Vehicles Activity of the ASM
Materials Systems and Design Division

Organized and edited by
LARRY J. VANDE WALLE
International Harvester Company

MATERIALS/METALWORKING TECHNOLOGY SERIES

American Society for Metals
Metals Park, Ohio 44073

PREFACE

The Technical Divisions of ASM were responsible for programming the 1979 Materials and Processing Congress. In particular, the Highway and Off-Highway Vehicles Activity of the Materials Systems and Design Division organized the four sessions covered by these proceedings.

The sessions were planned to expand on topic matter addressed in previous years or, as in one case, address a different topic. Composites, rare earth treated steel, nodular iron castings and aluminum castings were all discussed as they have been in the past; however, a few new concepts were involved. There were similar presentations on paint protection, surface preparation and zinc containing coatings. The new topic matter addressed at this Congress related to liquid retaining and sealing materials. Subject matter ranged from sealing porosity to gasketing to retaining fasteners.

Thanks are expressed to all of the authors, session chairmen and their companies. Without their willing dedication, the stage would have been empty. All compliments on these sessions received by the Technical Divisions must be reflected on these people. Special thanks also to the committee members for helping to organize these sessions. They are: F. E. Arant, the Timken Company; D. L. Cheever, Deere & Company; J. P. Faunce, Martin Marietta Laboratories; D. W. Gomersall, Inland Steel Research Laboratories; B. A. Juel, Funk Manufacturing; J. E. Miller, Delco Products Division; W. R. Phillips, Investigative Engineering Laboratories; R. H. Van Pelt, Caterpillar Tractor Company; and W. B. Young, Dana Corporation.

Thanks also to Elliot Boardman and Patti Gaeta at ASM headquarters for their support and assistance.

Larry J. Vande Walle
International Harvester Company
Committee Chairman

CONTENTS

DESIGN EFFICIENCY — I: HIGHWAY AND OFF-HIGHWAY VEHICLES

DESIGN EFFICIENCY — II: ADHESIVES AND SEALANTS IN THE STRUCTURAL WORLD

* Paper not available.

* Paper not available.

DESIGN EFFICIENCY — I:

HIGHWAY AND

OFF-HIGHWAY VEHICLES

High-Strength Reinforced Plastic Bumpers for Passenger Car Applications

N. A. HULL AND D. H. BERGSTROM
Ford Motor Company

Introduction

The growth over the past several years of the families of high strength reinforced plastics has both expanded the range of potential automotive structural applications and accelerated the pace at which these applications have progressed from the concept stage to production realities. The immediate benefits of such developments are obvious: weight savings accruing to plastic components contribute toward the goal of meeting legislated corporate average fuel economy requirements, and the cost savings versus other lightweight material candidates aid in the maintenance and improvement of the competitive position in the marketplace. The long range benefits are slightly more subtle: lower tooling costs allow less costly appearance changes, and the use of plastic to replace less abundant materials, such as aluminum, allows more flexibility in the allocation of material resources throughout the entire vehicle.

Ford Motor Company began, in late 1976, to investigate the potential of developing a high strength reinforced plastic bumper system as an alternative to conventional steel or aluminum assemblies. The basic objectives of the project were threefold:

- To develop a bumper which would be directly interchangeable with an existing production bumper.
- To be competitive with an existing aluminum bumper assembly on the basis of cost and weight.
- To meet all company bumper design and performance requirements as well as Federal bumper regulations (Part 581).

This paper presents the results of the above project and identifies major program issues where the combined expertise of the structural plastics industry is solicited to resolve these problems.

Background

The project described herein was the outgrowth of an earlier study of the production and performance feasibility of high-strength RP automotive bumpers. The initial study was not aimed at a particular carline application, but was

Reprinted from *Body Engineering*, Journal of the American Society of Body Engineers, Vol. 7, No. 1, Spring 1979.

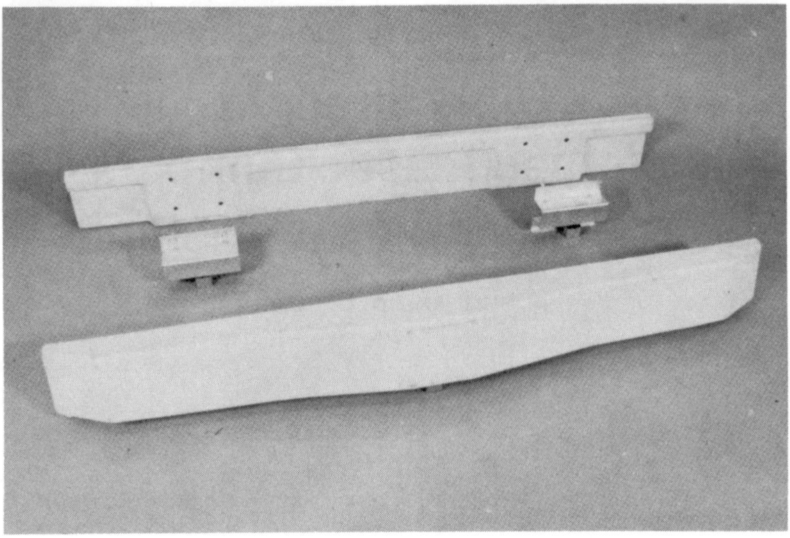

Fig. 1—Concept reinforced plastic bumper beam

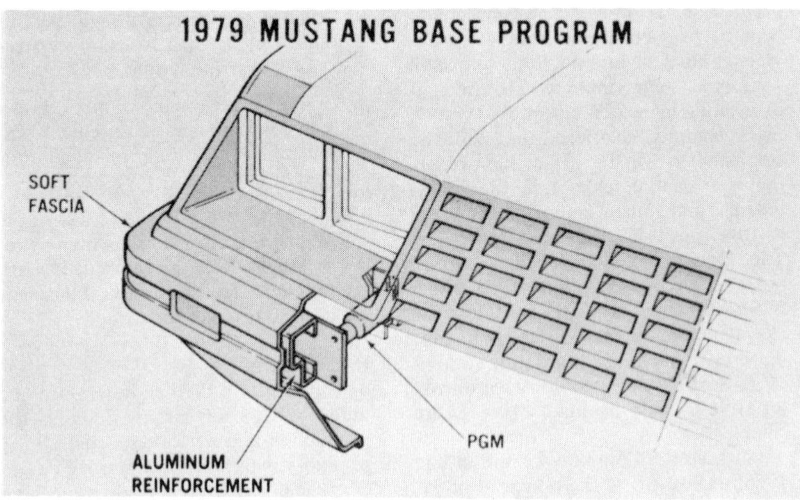

Fig. 2—1979 Mustang front end construction

intended to examine design and manufacturing parameters of a "typical" mid-size vehicle. The original design is shown in Figure 1, and consisted of a two-piece, bonded assembly with steel spacers at the mounting locations. Through impact testing at 5 mph in accordance with Part 581, a need for additional local strengthening of the beam assembly was identified. Combinations of bulkhead shape and location were evaluated, leading to a final design which, when tested, indicated satisfactory performance had been achieved. Accordingly, a program was established to refine the design and attempt application to a specific vehicle.

Design

The second phase of the RP bumper program was targeted at the 1979 Mustang, for several reasons. First, the vehicle was within the weight range for which the concept beam had been designed, and would also be typical of the weight of future downsized vehicles; second, representative vehicles would be available to support preliminary testing; and finally, the bumper would be hidden behind a reaction injection molded (RIM) urethane fascia, thereby removing any necessity for a highly finished surface on the RP component.

The 1979 Mustang front bumper system, as shown in Figure 2, consists of a soft, RIM urethane fascia attached to the body sheet metal. Behind the fascia, and mounted independently to the vehicle structure, is an energy-management system comprised of an aluminum reinforcement and hydraulic energy-attenuating devices. The high-strength RP bumper was designed to be a direct bolt-for-bolt replacement for the aluminum reinforcement, requiring no revisions to any other component of the bumper system.

The design which evolved from our analysis of the concept phase differed in two major areas: first, the bond joints were revised to both strengthen the assembly and to allow the rear plate to "cap" the front plate so as to prevent the front from "breathing" or trying to open during frontal impacts. Second, it was found that the multiple bulkheads in the concept beam could be replaced, at a cost and weight savings, by a single, "W"-shaped reinforcement. The design which was prototyped is shown in Figure 3. The bolt-on parking lamp brackets shown are aluminum, and are required so that the parking lamp may be serviced from the rear of the bumper, rather than having to remove the fascia.

Product Performance Requirements

In order to be acceptable for production implementation, the RP bumper (or any other bumper) must satisfy numerous product performance requirements, which may be categorized as:

- Federal regulations
- Company bumper design requirements
- Product Engineering requirements

The Federal regulations are outlined on Figure 4. These requirements were not in effect at the time the bumper was designed, but were anticipated to be enacted for the 1979 model year and beyond. The RP bumper was therefore designed to meet these more stringent requirements rather than safety standard FMVSS 215, which was applicable to pre-1979 passenger cars.

The Company bumper design requirements are in addition to the Federal regulations, and are listed in Figure 5. These requirements are Company practices established to provide products which will prove satisfactory through the normal anticipated usage of the vehicle.

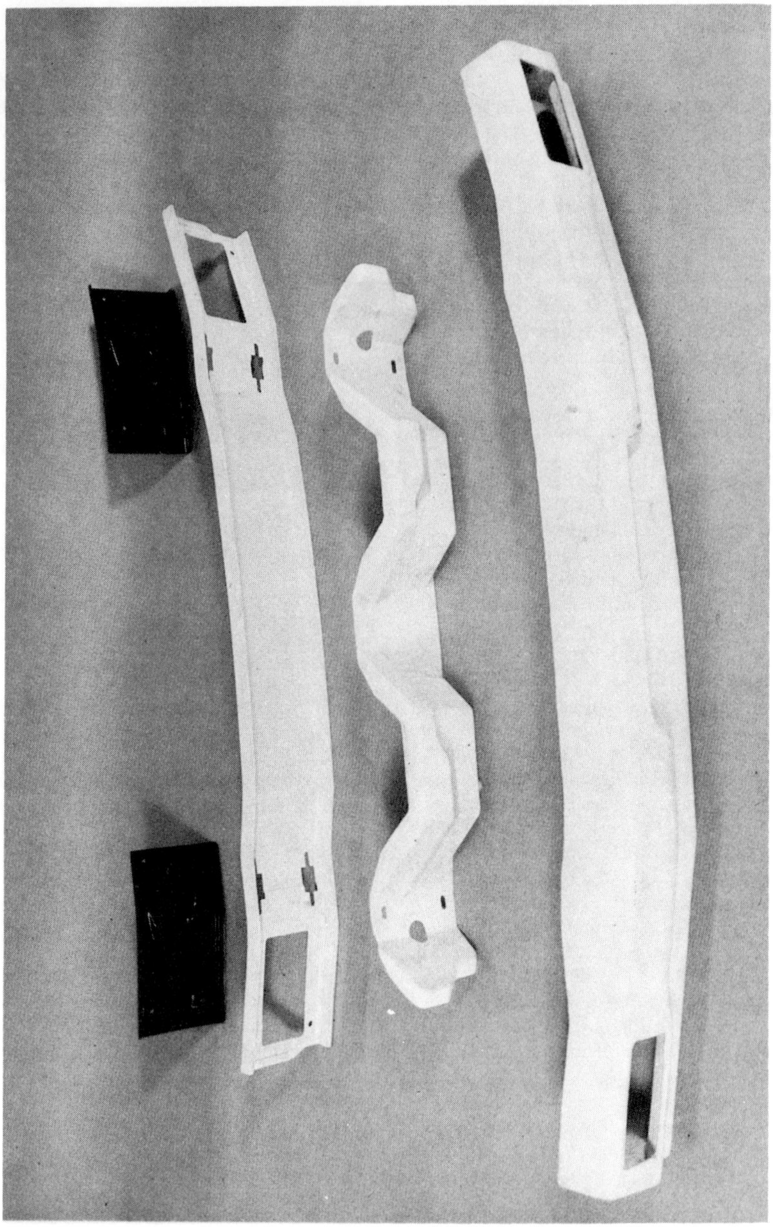

Fig. 3—Three piece reinforced plastic bumper for 1979 Mustang

The final category, Product Engineering requirements, are listed in Figure 6. They are the design requirements or targets which would be most involved in evaluating a candidate system and in choosing between design and material alternatives.

Material Development

From the results of the concept beam phase, it became evident that design was not the sole factor in developing a successful RP bumper. Crucial to the performance of the assembly was the selection of material and adhesive systems.

During the development of the Mustang bumper, three (3) adhesive systems (epoxy, modified acrylic, and polyurethane) and fourteen (14) material systems encompassing polyester, vinyl ester and epoxy resins were molded and evaluated. The results of physical property testing of typical systems are shown in Table I. Of particular interest is the wide range of values which occur in specimens taken from molded parts compared with supplier-published data and data from laboratory plaques. For example, the tensile strength measured on the molded parts, while displaying approximately the same mean strength as samples cut from laboratory plaques, had individual values in the longitudinal versus transverse directions which varied by a factor of two (2) to three (3). This extreme variation of physical strength measured in orthogonal directions on the molded parts highlights the difficulty involved in calculating the strength of a theoretical design, and reinforces the need for iterative design and testing programs. Our analysis of the parts led to the conclusion that the variation in strength was due to orientation of the reinforcements during the molding process. This was believed to be caused by both the shape of the parts and the inability of our temporary tooling to maintain adequate sealing to prevent material leakage.

Performance

The RP bumper assemblies were tested for compliance with Part 581, first at an outside laboratory for screening of candidate material systems, and secondly, at the Ford bumper test pendulum facility for verification of

PART 581 — BUMPERS

Impact Conditions

- Pendulum Impact
  - 3 MPH — 30° Corner - 2 impacts
  - 5 MPH — Longitudinal + 2 impacts
- Barrier — 5 MPH — 1 impact

Protection Criteria

- No damage to any exterior surface material, paint or polymeric coating
- No permanent deviation of bumper greater than 0.75 inch (19 mm) from original contour*
- No dent in bumper greater than 0.38 inch (10 mm) in contacted areas*
- No contact of test device (except impact ridge) in excess of 2000 lb. (907 kg)
- No leaks in fuel, cooling and exhaust systems
- Hood, trunk, doors, propulsion, suspension, steering and braking systems must remain in adjustment and operate normally
- Lighting devices free of cracks and comply with visibility requirements

**Effective September 1, 1979*

FIGURE 4

COMPANY DESIGN STANDARDS

- **JACKING**
- **TOWING**
- **RIDE AND HANDLING**
- **DURABILITY**

FIGURE 5

PRODUCT ENGINEERING CRITERIA

- **MANUFACTURING AND ASSEMBLY FEASIBILITY**
- **PRODUCT QUALITY AND RELIABILITY**
- **SERVICEABILITY**
- **COST/WEIGHT EFFECT ON VEHICLE**

FIGURE 6

TABLE I

TYPICAL HIGH-STRENGTH R.P. BUMPER MATERIAL PROPERTIES

MATERIAL (Reinforcement Level)	SUPPLIER DATA Tensile Strength (P.S.I.)	SUPPLIER DATA Flex Strength (P.S.I.)	SUPPLIER DATA Flex Mod (P.S.I. X 10^6)	TEST PLAQUE DATA Tensile Strength (P.S.I.)	TEST PLAQUE DATA Flex Strength (P.S.I.)	TEST PLAQUE DATA Flex Mod (P.S.I. X 10^6)	MOLDED BUMPER DATA Tensile Strength (P.S.I.)	MOLDED BUMPER DATA Flex Strength (P.S.I.)	MOLDED BUMPER DATA Flex Mod (P.S.I. X 10^6)
				L — Longitudinal T — Transverse			L — Longitudinal T — Transverse		
Vinyl Ester (50% Oriented)	Data are not published for oriented reinforcements			27,000 L 26,000 T	58,000 L 41,000 T	1.8 L 1.2 T	38,000 L 14,000 T	44,000 L	1.7 L
Polyester (65% Random)	31,000	44,000	1.6	26,000 L	41,000 L	1.4L	15,000 L 42,000 T	30,000 L	0.9 L
Epoxy (63% Random)	26,500	40,000	2.1	—	—	—	19,000 L 34,000 T	31,000 L	1.5 L

success. Although described previously as consisting of five (5) specific impacts, the test requirements were expanded for the screening portion of the program to encompass many more impact locations.

The original test plan for the Mustang RP bumpers was designed to evaluate the various candidate materials through a wide range of test conditions so as to identify viable combinations of materials and adhesives for the given design configuration.

This test plan was suspended when the initial tests indicated a good potential of meeting the performance targets on a cost and weight-effective basis versus the competing aluminum bumper. A decision was made to actively pursue production implementation on the 1979 Mustang and the test program was accordingly revised to focus on only one issue — meeting the original weight target.

The first RP bumpers to pass testing weighed 22 pounds (10.0 kg) versus a weight bogey of 14.5 pounds (6.6 kg). As a first step to achieving the weight target and qualifying the RP bumper for production, all material and adhesive evaluation was concluded and the leading system at that point in time was selected to "carry the banner" for the RP community. This system, in which all parts were molded from a 20% calcium carbonate filled, 50% glass vinyl ester compound and bonded with a two component urethane adhesive, was then subjected to a multi-faceted design analysis to attempt to meet the weight target.

Design Optimization

Attendant to the design issue of optimizing the weight of the RP bumper assembly was the constraint of meeting the new model timing schedule if the RP bumper were to be given serious consideration for production implementation. In the time remaining before a final commitment had to be made regarding selection of a bumper system, it was decided that only three approaches to weight reduction of the RP assembly would be investigated. These were:

- Eliminate the "W" reinforcement and fill the resulting 2-piece bumper with low density, rigid urethane foam to retain the beam strength properties.
- Create a new 2-piece design by combining the functions of the "W" and "back" into one part.
- Retain the original design and modify the temporary tooling to achieve minimum part weight.

The two-piece, foam-filled bumper assembly was the first to be evaluated. While showing nearly equivalent test results to the three-piece design, the substitution of foam for the "W" was unsuccessful in reducing the weight of the bumper assembly and was judged to require a greater development effort than the three-piece design. Therefore, it was eliminated from consideration for this program.

The "integrated W" two-piece design, as shown in Figure 7, was considered promising enough that a prototype tool was built to allow evaluation of the concept. The prototype parts were able to reduce the combined weight of the "W" and the "back" by 40 to 50%, which, in combination with weight-reduced "fronts," would have satisfied the overall weight target. However, the assemblies were unable to meet the requirements of Part 581. Due to material flow and reinforcement orientation during

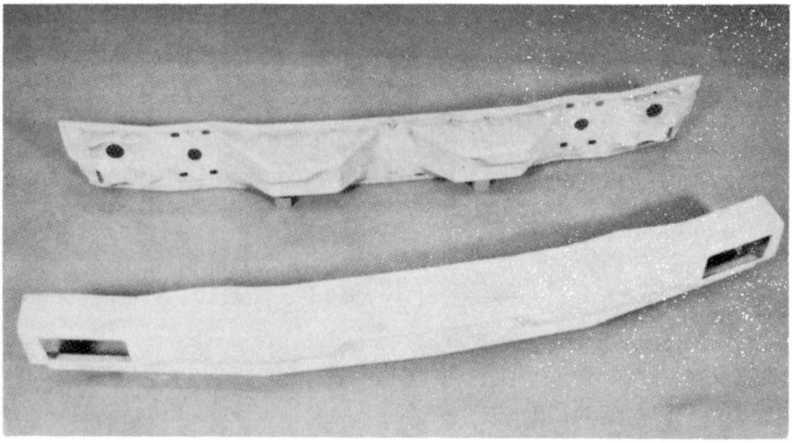

Fig. 7—"Integrated W" R.P. Bumper

molding, the "integrated W" lost all strength in the transverse (tensile) direction and fractured as the beam deflected during the pendulum testing, as shown in Figure 8. Subsequent attempts at strategic placement of continuous reinforcements in the transverse direction were unsuccessful, in that the continuous glass could not be retained in the required locations during the molding process. Although it was felt that the "integrated W" could have been made to perform, the time constraints of the program forced deferral of further effort on this concept.

The final approach to meeting the weight objective — modifying each of the three components to minimum weight within the limits of the prototype tooling — proved to be the most successful, as shown in Figure 9, from the initial assembly weight of 22 pounds (10 kg), we were able to build and successfully test RP bumpers in the range of 15 to 16 pounds (6.8 to 7.3 kg). this weight represented a 2 to 3 pound (0.9 to 1.4 kg) penalty over the aluminum assembly, thich had itself been weight-reduced by 1 pound (0.45 kg) over the same period of time.

A concluding test series was scheduled in order to increase the confidence of the releasing activity in the structural adequacy and repeatable assembly quality of the RP bumper. Five (5) identical bumper assemblies were subjected to the same typical Part 581 impact sequence; all five (5) assemblies passed the test without incurring any visible damage.

Cost and Weight

The 1979 Mustang has gone into production with both aluminum and steel bumpers. The bumpers can be selectively scheduled so as to achieve the desired vehicle weight on a specific model. That is, if a particular engine-body style combination has achieved its weight objective, a less expensive, but heavier, steel bumper may be used. If, however, the weight objective has not been attained, the lighter aluminum bumpers will be scheduled.

Having two bumper systems for the Mustang car line made more difficult the task of placing the RP bumper into production. First, time and manpower constraints would have made it difficult to add a third complete bumper into the production release system. For the RP

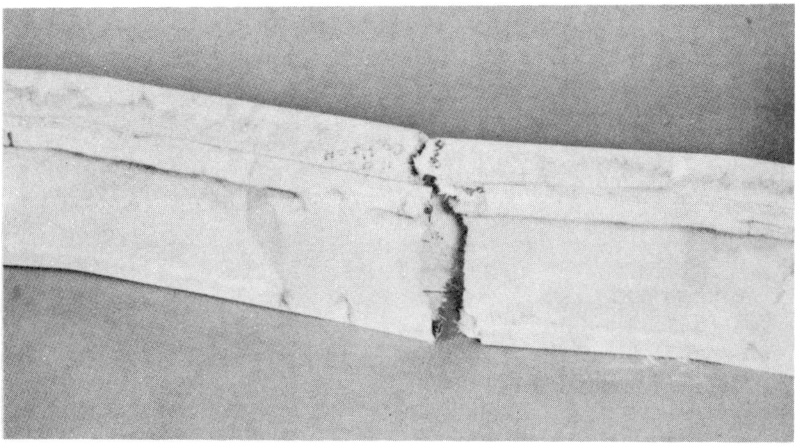

Fig. 8—Failed "integrated W" bumper

bumper to be selected, one of the metal alternatives would have to be dropped or substantial incremental engineering effort expended. Secondly, a similar concern was advanced regarding the investment both for development and for tooling which a third unique system would add to the 1979 Mustang program.

At the time the final bumper material selection was made, the relative cost and weight status of the three (3) candidate systems was:

	Aluminum	Steel	Reinforced Plastic
Cost	Base	38% Less than base	Base (Approx.)
Weight	13 lbs.	33 lbs.	15 to 16 lbs.
	(5.9 kg)	(15.0 kg)	(6.8 to 7.3 kg)

Because the steel bumper would be selectively scheduled for cost effectiveness and the RP bumper costs approximately 60% more than the steel, the only real choice was between the RP bumper and the aluminum version. Although the RP bumper, in a limited sample, showed promise of being able to satisfy the Federal regulations, there simply was not sufficient time remaining before new model introduction to resolve all the open issues and develop full confidence of being able to certify the RP bumper assembly for production. For this reason, the aluminum bumper was selected for the 1979 Mustang, and the development of the RP bumper was redirected toward future programs.

Conclusions

As a result of this development program, we have arrived at the following conclusions:

- A high-strength reinforced plastic automotive bumper is feasible from a design and performance standpoint.
- An RP bumper can be designed which will meet the applicable performance requirements within the same styling and packaging confines as a conventional metal bumper.

3-PIECE BUMPER

	BEFORE	AFTER
FRONT	10	8
BACK	3	2
"W"	3½	3
ADHESIVE	2	1
LAMP BRKT.	3½	2
	22 LB.	16 LB.

Fig. 9—Weight reduction of 3-piece R.P. Bumper

- An RP bumper can be designed which is both cost and weight competitive with other lightweight materials.

Our first effort at placing an RP bumper into production, while not accepted for implementation, has shown significant promise for future programs. There are many issues remaining — increased weight reduction, repeatability of manufacture and assembly, non-destructive test methods for assuring quality control and cost posture versus alternative lightweight materials, to name some of the more critical ones. With continued support and assistance from the structural plastics industry, we believe these issues will be resolved. We are equally confident that these solutions will permit us to add bumpers to the growing list of products where reinforced plastics technology has proved successful in the search for lighter, more fuel-efficient vehicles. ☆

Hybrid Graphite Fiber Composites for the Automotive Industry

S. L. Cross
Hercules Inc.

ABSTRACT

This paper summarizes the status of hybrid graphite fiber composites as related to the automotive industry. Included are: (1) basic raw material forms available for use in hybrid construction, (2) relative tensile properties of unidirectional composites as compared with hybrid forms being developed, (3) some example applications, with emphasis on raw materials and construction, and (4) projected economics for the potential use of hybrid materials. Over the past several years, raw material suppliers have concentrated on developing material forms suitable for high-volume manufacture of automotive parts. Emphasis has been on the more economical hybrid construction. In general, hybrid graphite construction includes between 5 and 25% graphite reinforcement, with the balance being fiberglass and resin. Future economics based on projected escalation curves look favorable for hybrid graphite composites as a result of anticipated reduced fiber prices.

HYBRID GRAPHITE FIBER COMPOSITES have received much attention from the automotive industry over the past several years. While this initial exposure has generated interest by virtue of the potential weight reduction, many pertinent issues have been raised concerning the use of graphite fiber. Issues such as cost effectivity, availability, producibility, and long-term durability are some of the key questions to be answered before high-volume use will be considered. In view of these requirements, material and component suppliers are actively pursuing: (1) new material forms for high-volume production, (2) a more extensive design data base, (3) a better definition of processes required to achieve production rates, (4) refinements in component designs to yield cost-effective construction, and (5) extended test programs of prototype parts to establish service life characteristics.

Requirements placed upon the Transportation Industry to become more fuel efficient have served as a catalyst for investigation of new, lighter-weight materials. Not only has this brought about the examination of fiber-reinforced composites, but it has spurred more efficient use of metals. Vehicle downsizing, more efficient engine and drive line systems, and aerodynamically cleaner body designs are all providing significant improvements in reduced energy consumption. Within the materials category, plastics and fiberglass composites offer direct savings in weight, as well as cost, and represent materials of familiarity to the auto industry. Hybrid graphite composites probably offer the greatest potential weight reduction approach; however, they also present the greatest challenge in meeting cost constraints and producibility. No doubt these materials are viewed as a "final approach" and will find their place in selective structural applications as more and more weight reduction is required. Like any change involving new and untried materials, it will take time to resolve issues and gain the needed confidence. This will only happen through continued industry backing of development and test programs. The major goals are, therefore, to develop the required material forms and necessary data base and to demonstrate the required manufacturing processes and economics for its use.

RAW MATERIAL FORMS

Graphite fiber materials available today fill a broad matrix of forms from which hybrid composites are constructed. A basic fiber (Table 1) can be selected having high strength (400 to 450 ksi) and a typical modulus of 30-34 ksi or high modulus (50-55 Msi) and strengths of 300-350 Msi. The high-modulus fiber is currently much more expensive, but future projec-

Table 1 - Fiber Types and Bundle Size

	Strength (ksi)	Modulus (Msi)	Density (lb/in.^{3})
High strength	400-450	30-35	0.064
High modulus	300-350	50-55	0.067

Bundle Size	Application
3,000	Fabric
6,000	Fabric, XMC
10,000	Prepreg
12,000	Prepreg, SMC, F.W.
40,000	SMC, Pultrusion, F.W.

tions indicate prices at more competitive levels.

Graphite fibers (Table 1) are generally available in bundle sizes (filaments per tow) from 3000 (highest price) to 40,000 (lowest price). This latter fiber has only recently become available and has been developed primarily for the automotive market. The 3000 and 6000 filament tows are generally used in the manufacture of woven fabrics, although 6000 filament tow is being considered in the manufacture of hybrid XMC*. The more standard 10,000 filament tow is being used in hybrid SMC, pultrusion, and filament winding.

*PPG trademark.

CONTINUOUS	CHOPPED
SPOOL, BEAM, BOX	MILLED 1/4 MM
SIZED, UNSIZED	1/8 IN., 1/4 IN.,
HIGH STRENGTH, HIGH MODULUS	1-2 IN. SIZED
FILAMENT WINDING	INJECTION MOLDING
PULTRUSION. PREPREG	

Fig. 1 - Graphite fiber products

In addition to type and bundle size, the fibers are available in various forms (Fig. 1 and 2) such as continuous, chopped, woven, and resin-impregnated. These forms have been developed and tailored to fit various processes for final part manufacture. Continuous fiber, packaged on individual spools or in boxes, offers a lower-cost approach to part manufacture through filament winding and pultrusion because no intermediate conversion costs related to a material form are needed. Sized fiber and beamed fiber are used to manufacture sheet type molding compounds for compression

WOVEN, KNIT, MAT

BIDIRECTIONAL
UNIDIRECTIONAL
RANDOM
MFG. COMPLEX
CONFIGURATIONS

RESIN IMPREGNATED

UNIDIRECTIONAL, MULTI-DIRECTIONAL
EPOXY, POLYESTER, VINYLESTER
1/4 IN. TO 48 IN. WIDE
PART MANUFACTURE
TAPE WRAP, LAYUP

Fig. 2 - Graphite fabric and prepreg product forms

molding. This form does require intermediate conversion costs; however, projections for high volumes indicate potential \$.50 per pound conversions, which is low compared with that required for the fabrics and resin-impregnated tapes shown in Fig. 2. These products find wider acceptance in the aerospace and leisure markets primarily because of the more specialized products required by these markets.

BASIC PROPERTIES

Programs are underway to characterize new hybrid material forms for the automotive industry. The all-glass, all-Kevlar, and all-graphite structural composites have been well characterized through aerospace and industrial programs. Studies are now focusing on hybrid forms of SMC, XMC, and injection-molding compounds. The key tradeoff involves properties and economics.

In the design of prop shafts, leaf springs, control arms, various torsion rods, crossmembers, etc., basic composite properties (Table 2) are considered the building blocks to final construction. While numerous other properties are required for each design case (such as compression, shear, fatigue, impact), only tensile strength, modulus, and density are presented for this comparison. Because the various component designs are tailored through fiber selection and orientation to achieve maximum efficiency and minimum cost, several orientations are presented to illustrate the available choices. A particular design, such as a hybrid composite prop shaft, might use unidirectional graphite fiber and multidirection ($\pm$ 45°) fiberglass in combination. The

Table 2 - Basic Composite Properties

		Tensile Strength ksi (MPa)	Tensile Modulus Msi (GPa)	Density lb/in.^{3} (g/cm^3)
Uni-dir (0°)	graphite	220.0 (1517)	20.0 (138)	0.056 (1.55)
	fiberglass	160.0 (1104)	5.6 (39)	0.072 (1.99)
Bi-dir (0,90)	graphite	122.0 (838)	10.7 (74)	0.056
	fiberglass	75.0 (518)	3.6 (25)	0.072
Multi-dir (±45°)	graphite	20.0 (138)	2.5 (17	0.056
	fiberglass	22.0 (152)	1.6 (11)	0.072
Isotropic (0, ± 45, 90)				
	graphite	88.0 (604)	6.9 (48)	0.056
	fiberglass	48.0 (330)	2.6 (18)	0.072

*Nominal 60% fiber volume, epoxy resin

graphite provides bending stiffness, while the glass is designed to carry the torsional loading. The numerous possibilities for tailoring a specific construction are obvious.

The influence of graphite fiber addition to all-fiberglass construction can be seen in Table 3 for unidirectional combinations. These examples assume intraply mixing of fibers where the strengths are based on graphite fiber strain limitation. The all-fiberglass and all-graphite values are less than those given in Table 2 due to lower fiber volumes and the resin systems used. (Values in Table 2 reflect higher performance epoxy resins.) Improvements in fiber loading and compression moldable polyester will increase the values shown. The primary advantage of graphite fiber addition involves improved stiffness -- a 40% addition of graphite fiber increases stiffness by 96%. Recognition of this trend is important when making initial material selections for preliminary design analysis.

Table 3 - Unidirectional Graphite/Fiberglass Hybrids

	Tensile Strength ksi (MPa)	Tensile Modulus Msi (GPa)	Density lb/in.^{3} (g/cm^3)
All fiberglass	130 (897)	5.6 (39)	0.072 (1.99)
Graphite/fiberglass			
20/80	87.4 (603)	8.3 (57)	0.069 (1.91
40/60	116.0 (800)	11.0 (76)	0.066 (1.82)
60/40	143.0 (986)	13.6 (94)	0.062 (1.71)
80/20	172.0 (1186)	16.3 (113)	0.059 (1.63)
All graphite	200.0 (1379)	19.0 (131)	0.056 (1.55)

*Nominal 50% fiber volume, polyester resin

Interply hybrids, which are representative of SMC and XMC compression molding compounds, are similar to a sandwich construction. This includes combinations of chopped glass and continuous glass or graphite fiber. Several examples are summarized in Table 4. As noted previously, work is continuing in this area to provide better material characterizations and economics to establish tradeoff benefits. The first three molding compound materials shown provide a base reference to all-glass construction. By replacing a 40% continuous, 20% random fiberglass material with a 12% continuous graphite, 48% random fiberglass construction, strength is increased by 60% and stiffness is increased by nearly 90%. As expected, relatively small additions of graphite fiber reinforcement yield proportionally greater increases in strength and stiffness. The first three hybrid combinations are more indicative of the SMC process, while the fourth represents an XMC type material with continuous fiberglass core and graphite outer layers.

Table 4 - Molding Compound Materials

	Tensile Strength ksi (MPa)	Tensile Modulus Msi (GPa)	Density lb/in.^{3} (g/cm^3)
(Molding compound sheet)			
All fiberglass			
Random 60%	30 (207)	2.5 (17.3)	0.072 (1.99)
Continuous (unidir) 60%	70 (483)	5.5 (38.0)	0.072 (1.99)
Continuous (40%) random (20%)	50 (345)	4.0 (27.7)	0.072 (1.99)
Graphite/fiberglass			
Graphite (12%) random glass (48%)	80 (552)	7.5 (51.9)	0.069 (1.91)
Graphite (15%) random glass (45%)	90 (621)	8.0 (55.3)	0.068 (1.88)
Graphite (20%) random glass (40%)	100 (690)	9.5 (65.70	0.067 (1.85)
Graphite (20%) cont ±15° glass (40%)	120 (827)	10.5 (72.6)	0.067 (1.85)
All graphite			
Continuous (unidir) 60%	190 (1310)	16.0 (110.6)	0.056 (1.55)

Note: Assumed construction, graphite face sheets - fiberglass core

These are just a few of the forms being examined today to identify potential products for the automotive industry where economics will be established as a function of material characteristics and processibility. Other materials being considered include injection molding compounds with various combinations of fiberglass and graphite reinforcement, improved property thermoplastic sheet material for hot forming or stamping, and reinforced reaction injection molding (RIM) material.

APPLICATIONS AND RAW MATERIALS

As previously discussed, there are many forms of raw materials and many possibilities for combining these materials to achieve properties tailored to each particular application. For several years, applications have been under development with these materials to examine potential weight savings and economic feasibility. Several of these applications are summarized relative to the raw material used, basic construction, and manufacturing process. This section addresses relative material costs, potential weight savings, and potential for future economic payoff.

While 20 to 30 applications with hybrid graphite composite construction are being evaluated today, many issues must be answered before these will reach limited production status. Figures 3 through 7 illustrate several of the applications being considered. In

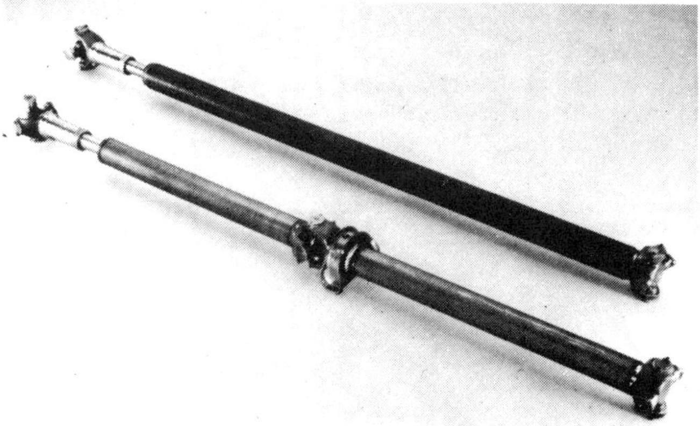

	HYBRID GL/GR	HYBRID GR/METAL
MATERIAL:	HIGH STR GRAPHITE/ E-GLASS/METAL	HIGH MOD. GR./Al
CONSTRUCTION:	14% GR/38% FI/21% RES/27% MET.	6% GR/4% RES/90% Al
PROCESS:	FILAMENT WIND	TAPE WRAP

Fig. 3 - Composite prop shaft

MATERIAL:	HIGH STR. GRAPHITE/E-GLASS/METAL
CONSTRUCTION:	16% GR/26% GL/28% RES/30% METAL
PROCESS:	COMPRESSION MOLD

Fig. 4 - Hybrid leaf spring

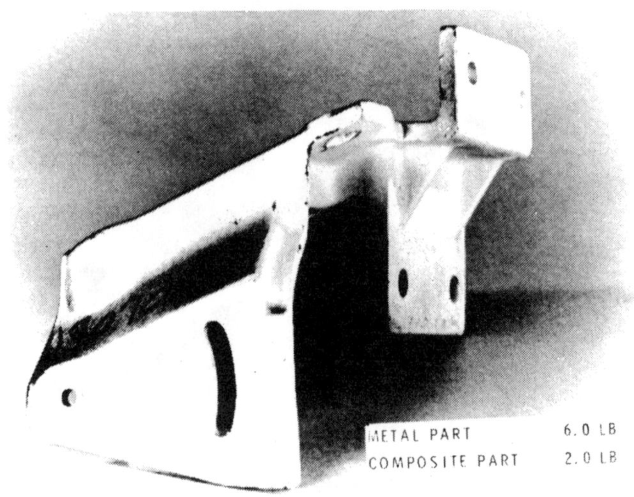

MATERIAL:	HIGH STR. GRAPHITE/E-GLASS
CONSTRUCTION:	27% GR/43% GL/30% RESIN
PROCESS:	COMPRESSION MOLD

Fig. 5 - Engine bracket

MATERIAL:	HIGH STR. GRAPHITE/E-GLASS
CONSTRUCTION:	5% GR/45% GL/50% RESIN
PROCESS:	PULTRUSION

Fig. 6 - Composite bumper

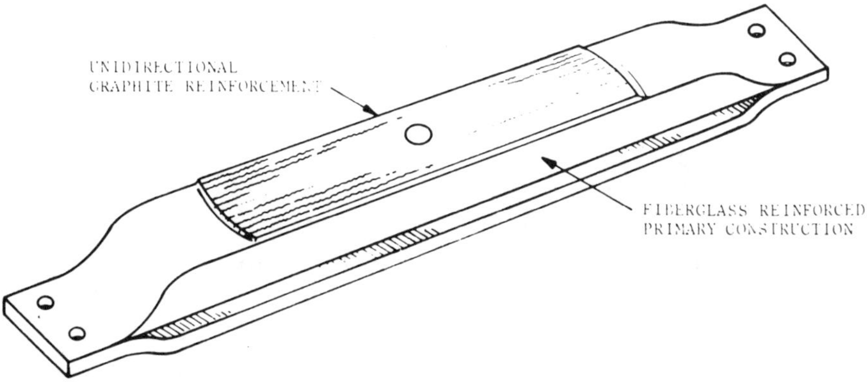

Fig. 7 - Hybrid transmission support

general, graphite fiber content is below 20% by weight to minimize material cost. Further, the high-strength fiber is used primarily because it is the lower cost material and it offers significant improvements in stiffness (as well as strength) over the more commonly used fiberglass.

As low-cost, high-modulus fibers are developed, more use will be made of these fibers where the estimated cost delta is 20% over high-strength fiber. Since graphite fiber is used primarily for stiffness, the higher modulus fiber will yield lower raw material costs (60% higher stiffness than the high-strength fiber).

Filament winding, compression molding, and pultrusion have all shown a potential for high-volume manufacture. Production winding machines have been designed to produce over 30 prop shafts per hour. Multiple machines could double or triple this rate. Automated compression molding lines are being considered to reduce the current 3 minute mold cycle time to less than 1 minute. Pultrusion lines currently run at 75 feet per hour with relatively complicated shapes. Faster speeds with multiple machines and multicavity dies could produce sufficient stock material to satisfy production demands.

Two approaches are being considered for the prop shaft design (Fig. 3) -- the graphite/fiberglass hybrid and the graphite/aluminum (or steel) hybrid. The first approach includes high-strength graphite fiber in combination with E-glass fiber. These shafts are manufactured by the filament-winding process using multiple spools of continuous fiber. Metal sleeves are wound in to provide the interface for welding to yokes. The majority of material is low-cost fiberglass. Resin is applied during the winding process to minimize conversion costs. The overwrapped metal approach currently uses high-modulus fiber in resin-impregnated tape form. While the majority of material is aluminum or steel, the cost of resin-impregnated tape makes this approach less economical. Another approach would be to machine-lay the fibers in a wet-lay process, thereby eliminating the more costly prepregging step.

Several approaches are being considered for the manufacture of leaf springs (Fig. 4). These include compression molding, filament winding, and pultrusion. The example shown uses compression molding of glass/graphite resin-impregnated tapes. For the heavy truck applications, a hybrid composite spring pack is combined with a steel top leaf for attachment purposes and economics. By virtue of the composite tape form used, these assemblies are yielding high premium levels. Further work must be done with lower-cost hybrid molding compounds having shorter processing cycle times to achieve the desired lower premiums.

The rather complicated bracket configuration (Fig. 5) is compression molded with a form of sheet molding compound developed by Armco Composites. The potentially low conversion cost to produce this material form, which uses chopped fiberglass and continuous graphite fiber (sized with a polyester compatible resin and beamed to a 12 inch width), makes this approach economically attractive. While mold cycle time is approximately 3 minutes, improved molding compound resins are being considered to reduce this cycle time. For production, multiple dies with semiautomated material feed lines can provide the means for achieving high production rates. This single application is expected to demonstrate numerous other bracket type applications by virtue of the potentially low premiums and high weight savings over cast metal parts.

While the pultruded bumper (Fig. 6) has had difficulties in maintaining fiber placement during the pultrusion process, it does represent a viable application and processing approach in achieving lightweight structural beam components. This application uses a minimum amount of continuous graphite fiber to achieve the desired beam bending stiffness. The part is designed for 5 mph impact standards. Other beams having similar impact functions are being investigated for the pultrusion process. In some cases, all-glass construction may be required to achieve the desired higher deformation strain energy more typical of a ductile metal. The tradeoff is apparent when considering required strength versus required strain energy. A second approach using the compression molding process of sheet molding compounds is being considered for bumper backup beams.

The final example shown (Fig. 7) includes a transmission support crossmember which uses a small amount of continuous graphite fiber to achieve the required spring rate in combination with continuous glass XMC for compression molding. This application has the potential for weight savings in excess of 70% over the current steel part and a projected premium of less than $1.00 per pound of weight saved. The fiberglass meets gross beam bending requirements as well as critical shear stresses generated from torsional loads.

FUTURE ECONOMICS

The automotive industry has claimed for several years that graphite fiber prices must fall below $10/lb to be economically feasible. Fiber price forecasts since 1976 have indicated prices will fall below this level when volume exceeds one million pounds per year. This forecast trend is illustrated in Fig. 8; estimated timing when the volumes could occur is shown in Fig. 9. These figures do not reflect the aerospace market, which should exceed the

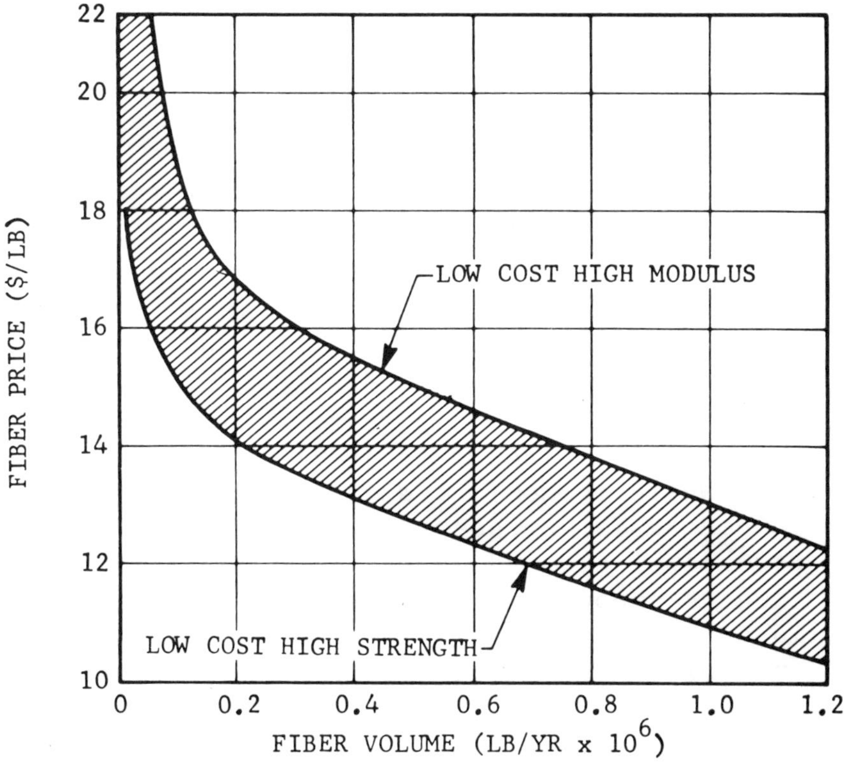

Fig. 8 - Fiber price/volume

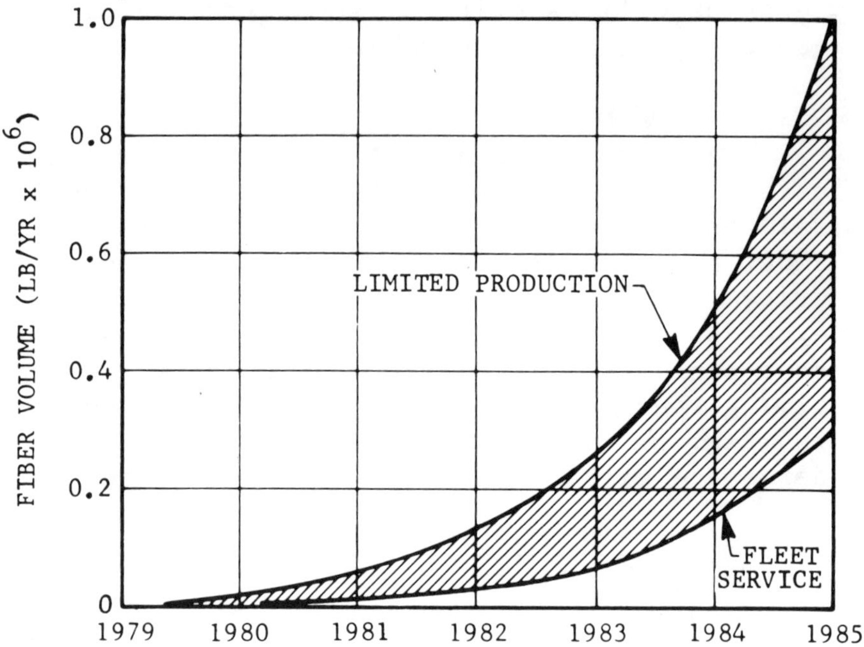

Fig. 9 - Potential volume/year

million pound level by 1981-1982 primarily for higher priced fiber products. This trend reflects a best estimate for the automotive market and the lower cost commercial graphite fibers and is based on several limited production programs occurring in 1981-1983.

While economic feasibility of any given component is dependent upon raw material cost, producibility, and added value for weight reduction, the potential unit cost of raw material must be reviewed as a function of construction and volume to establish baselines for material selection. Fig. 10 illustrates the potential unit costs of fibers, steel, and aluminum. Three hybrid graphite constructions are con-

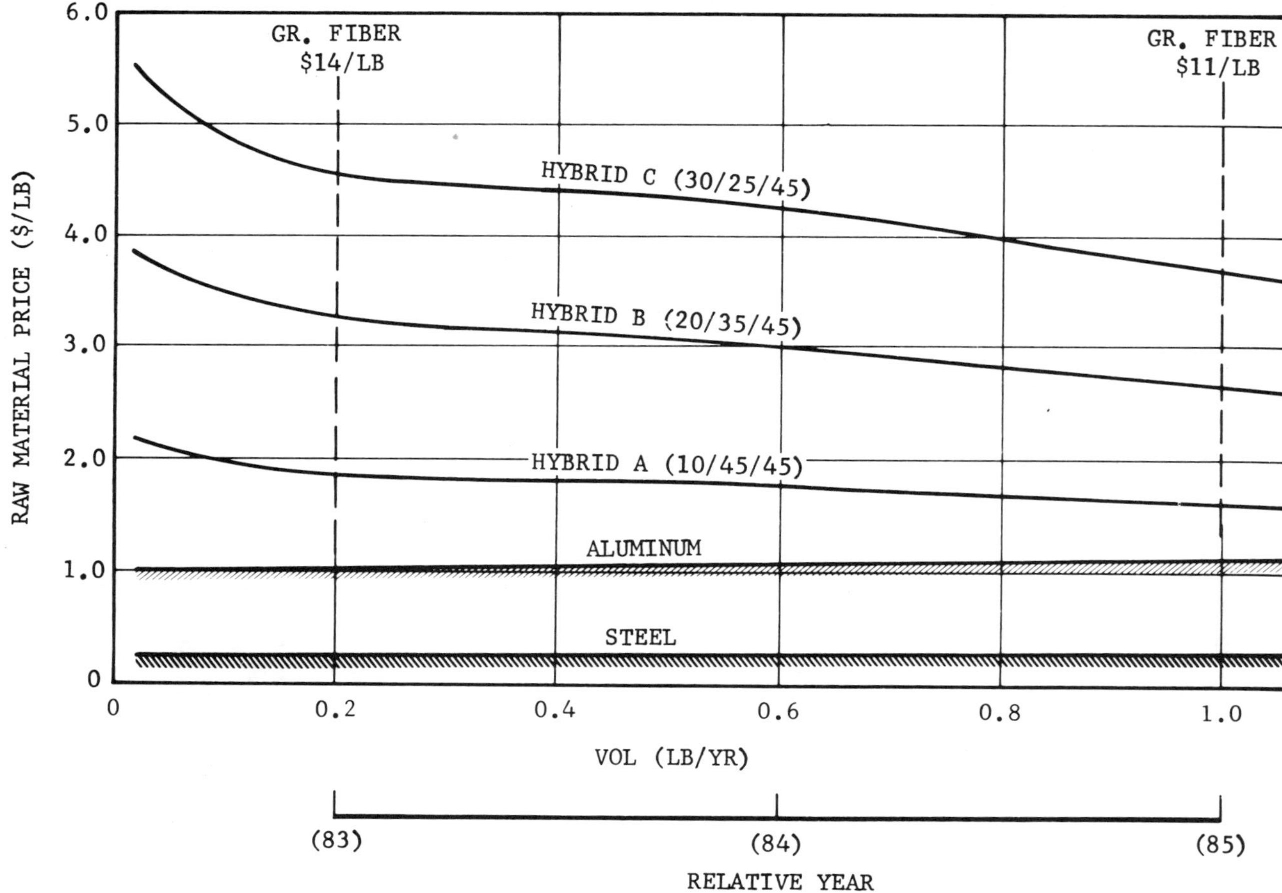

Fig. 10 - Raw material price

sidered: 10, 20, and 30% graphite (balance includes % glass, % resin). As shown, unit price of raw material runs from $1.80 - $3.80 per pound at the one million pound volume of graphite fiber which is estimated for 1985. On a cost per unit volume basis, these prices are $.12 - $.24 per cubic inch. Considering the reduced weight of components (50-70% weight reductions), a $2.0 - $4.0 per pound raw material comes very close to being economically competitive with metal construction at less than $1.00 per pound premium on weight saved.

Another important factor is the long-term effect of escalation. While the percent of graphite used in the examples shown runs from 5% to 30%, graphite fiber is responsible for 20-60% of the cost of the part under current low-volume projections. However, allowing for higher volumes and escalation of all other material and labor cost through 1985, for example, the relative cost of the graphite drops to 10-40%. Since the major cost component (graphite fiber) is dropping in price, the net effect is an overall lower escalation. An example of this projection is shown in Fig. 11 for the hybrid composite prop shaft compared with a two-piece, all-metal part.

While a big factor for this example is the removal of the center bearing assembly, this trend reveals the potential for direct cost savings (no premium) and a weight savings in excess of 50%. The spread given for the two-piece steel reflects minimum price for highest volume sizes, while the upper price reflects the lower volume sizes. The spread for hybrid composite shafts reflects minimum and maximum use of graphite fiber. The upper limit includes graphite fiber at a constant $18 per pound; the lower limit includes graphite as a function of projected volume. As shown in Fig. 11, a crossover in cost is projected when the potential production rate of 100,000 to 200,000 prop shafts occurs.

Although most applications considered as examples here show no crossover trend prior to 1985, they do show the reduced rate of

GRAPHITE FIBER PRICE		
1979	$18/lb	5000 lb
1981	$16/lb	50,000 lb
1983	$14/lb	200,000 lb
1985	$11/lb	1,000,000 lb

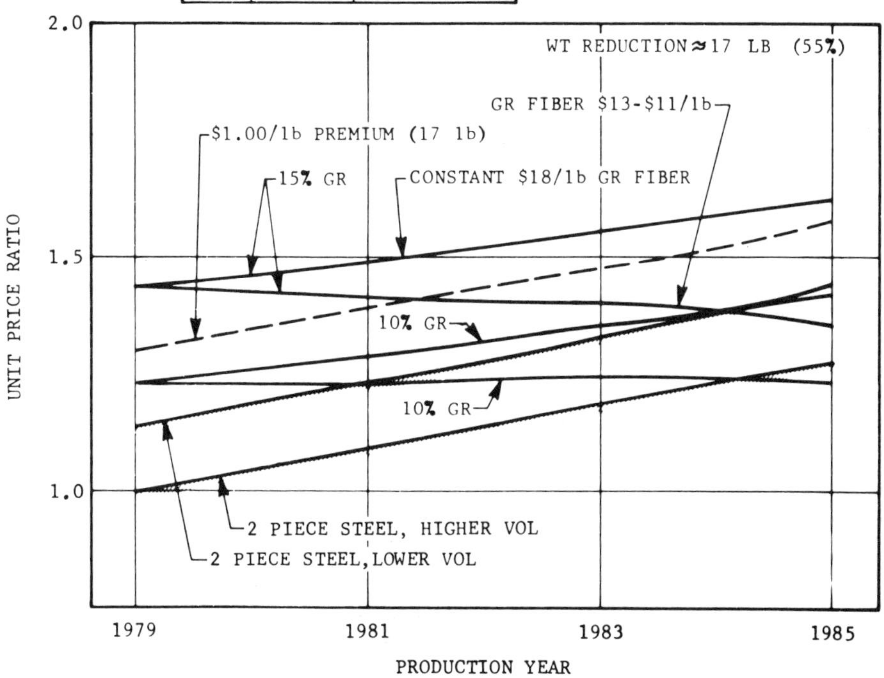

Fig. 11 - Prop shaft economic potential

escalation (Fig. 12, 13, and 14), which yields a potential for reduced premium on weight saved (under \$1.00 per pound). For the bracket (Fig.12), a \$1.00 premium is projected for 1985 (one million pounds fiber), after which the premium drops to around \$0.60 per pound saved. The upper limit reflects a slow market growth for which the fiber price remains at \$18 per pound.

For the transmission support crossmember (Fig. 13), the potential weight savings is 16 pounds out of the total 22 for the metal component. This is an ideal situation for hybrid composite construction because it can be tailored to satisfy local stiffness constraints

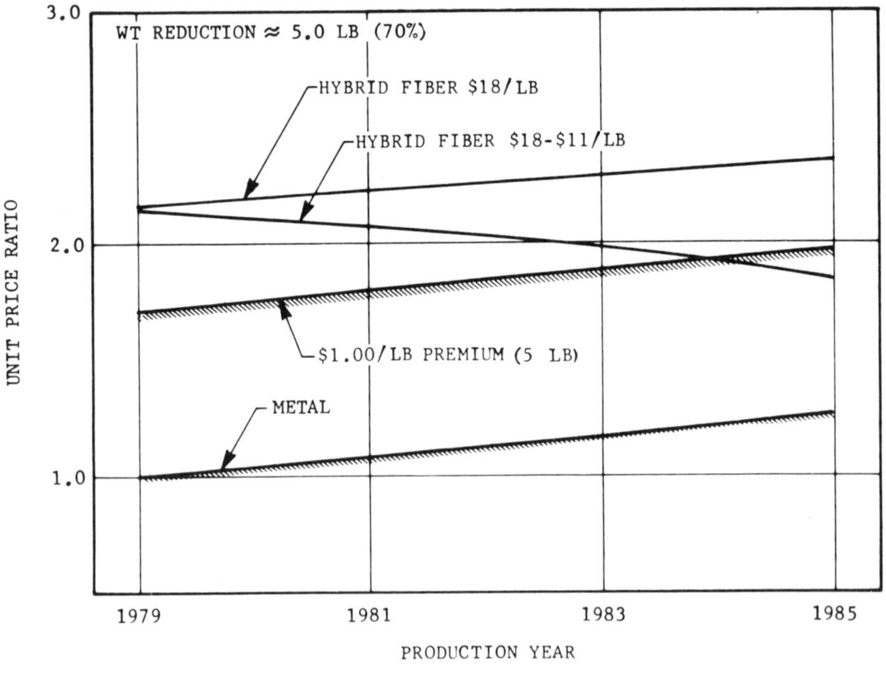

Fig. 12 - Bracket economic potential

(localized spring rate at transmission attach point) without penalizing the total crossmember length with unnecessary material. The metal part uses a standard gage steel tube of a thickness required for the localized stiffness. The hybrid design using 5% graphite reinforcement has across-the-board potential for under \$1.00 per pound premium on weight saved, while the 10% graphite hybrid does not fall below the \$1.00 per pound premium until volume reaches the estimated production level in 1983. This application assumes the use of graphite/ polyester prepreg tape in combination with continuous glass polyester molding compound where the price of graphite prepeg starts at

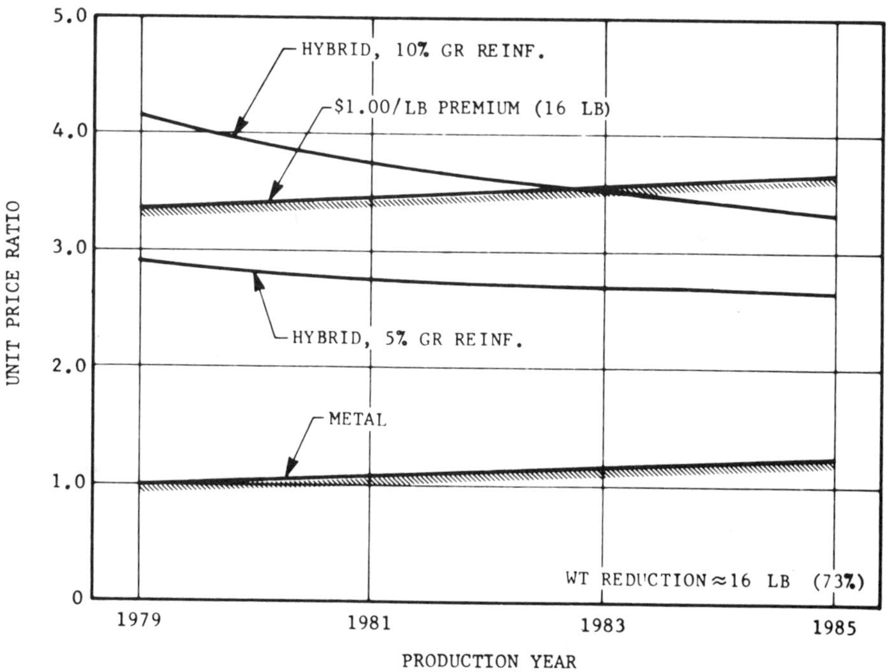

Fig. 13 - Crossmember potential economics

$30 per pound and decreases to $16 per pound by 1985.

The hybrid graphite leaf spring (Fig. 14) shows no strong potential for low premium on weight saved for the current design (max graphite) or processing method (compression

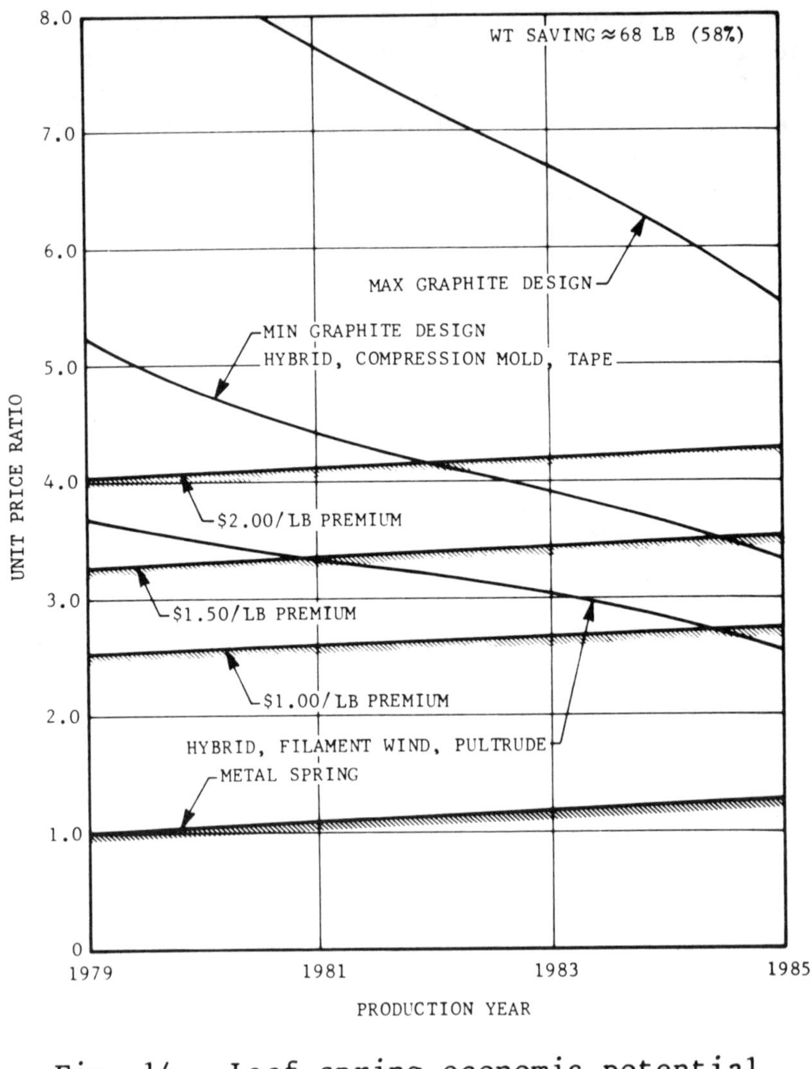

Fig. 14 - Leaf spring economic potential

molding with prepreg tape). If the current graphite prepreg tape requirement can be reduced from 13 pounds to 7.0 pounds (minimum graphite design), the forecast premium is reduced significantly. However, unlike the other three examples which include estimated manufacturing costs, this comparison only includes raw material prices. Therefore, the premiums shown will increase proportionally by the manufacturing costs. If the springs can be produced by filament winding or pultrusion, the potential for an acceptable premium level is much greater because conversion costs to prepreg tape are not required.

These applications are typical of the areas being investigated today. Hercules recognizes that current costs discourage many within the industry from commiting to programs with specific target production dates. However, for these examples we have not considered a potential for secondary weight savings based upon the primary or direct weight savings. In many cases, a direct component weight reduction can lead to further reductions in adjacent or other supporting components. Further, no consideration has been given to the value of weight reduction relative to potential CAFE penalty for not meeting MPG requirements due to excess vehicle inertial weight. Nor has the value of reduced life cycle costs relative to reduced fuel consumption and potentially higher fuel costs been evaluated. In the future, as higher gas mileage is required and as fuel costs increase, the overall added value for weight reduction will have more significance and may well justify the premiums forecast at this time.

SUMMARY

Hybrid graphite fiber construction is currently viewed as a final approach to maximum weight reduction. However, as the issues are resolved and better understanding and higher confidence in its use is achieved, wider acceptance is anticipated. Continued development and refinements in the raw material forms will lead to more efficient and economical introduction into currently used fiber-reinforced molding materials. More complete characterizations of the properties and processibility of these hybrid materials will provide better options for the material selection. Basic data available today have shown the potential for improved structural characteristics. Further studies are required to fully assess the economic potential. Automotive industry interest and support are vital to the success of this new material. Until this support develops, the acceptance and increasing use by aerospace, industrial, and leisure industries will ensure continuing growth and raw material price reductions. However, the automotive industry clearly represents the greatest potential for high-volume consumption and, therefore, the greatest driving force for achieving the necessary low-cost material.

NON-DIRECTIONAL, HIGH STRENGTH, HIGH TOUGHNESS STEEL FOR AGRICULTURAL IMPLEMENT DISKS

Laxmidhar Jena and G. H. Walter
International Harvester Co., Science and Technology
7 South 600 County Line Road
Hinsdale, Illinois 60521

ABSTRACT

A high strength, high toughness steel with uniform fracture toughness parallel and perpendicular to the rolling direction has been developed. This was produced by treating AISI 1085 steel with rare earth (RE) additions to produce RE sulphides which do not deform during the rolling operations. In the hardness range of RC 40-44, in both austempered (bainite) and quench and tempered (martensite) conditions, the steel was evaluated for fracture toughness parallel and transverse to the rolling direction using "tapered double cantilever beam (TDCB)" specimens. Fracture toughness parallel to the rolling direction increased with increasing proportion of globular sulphides. The sulphide-morphology-controlled 1085 steel was found to be superior to conventional straight rolled 1085 steel, but more importantly it was equal to cross-rolled 1085 steel with regard to fracture toughness isotropy.

Also austempered, and quenched and tempered conditions were evaluated using "dynamic tear (DT) type" blunt-notched (ground notch) and fatigue-precracked 3-point bend specimens. The austempered RE treated steel out-performed the quench and tempered RE treated steel with regard to notch toughness measured with ground-notched specimens. Quench and tempered steel performed better than austempered steel with regard to toughness when measured with fatigue precracked specimens.

The ground notch test results correlated well with ball tests (a production quality control test) and with controlled field tests conducted under severe impact conditions. Fatigue precrack test results did not appear to correlate with ball test or field test results.

The steel has been successfully used for the production of superior quality, heavy duty harrow disks (U.S. Patent No. 4098 622) eliminating the need for expensive cross-rolling to minimize directionality.

INTRODUCTION

Microscopic examination of field returned harrow disks, which had fractured in a straight line manner, indicated that MnS inclusion stringers contributed to the straight line fracture in the rolling direction. In order to eliminate MnS inclusion stringers and to produce non-deformable globular sulphides in disk steel (AISI 1085 grade), experimental sulphide-morphology-controlled steel samples were produced in co-operation with Wisconsin Steel.

The principle of sulphide morphology control has been successfully applied to the production of high-strength low-alloy steels(1). This concept of sulphide shape control is based on the comparative values of standard free energy of formation of metal sulphides.

Fig. 1 shows the standard free energy of formation of a number of sulphides(2). Sulphides of Ti, Mg, Zr, Ca and rare earth metals have lower free energy of formation than MnS, and form more stable sulphides than MnS. Because of the very high melting point of these sulphides they do not soften at steel rolling temperatures and maintain their globular shape in highly rolled products, see Table 1(3). In the presence of sufficient amounts of these elements in steel the deformable MnS inclusions do not form.

Experimental disk steel samples were made by treating AISI 1085 steel with Ti, Ca (Hypercal), rare earth silicide, and mischmetal (rare earth). Tables 2 and 3 give the composition of "Hypercal", rare earth silicide and mischmetal(2). Our trials with Ti, Ca and rare earth silicide did not produce 100% non-deformable globular sulphides.

The rare earth (mischmetal) treated 1085 steel was evaluated for fracture toughness using "tapered double cantilever beam (TDCB)" specimens(4) and compared with commercial grade 1085 steel and disk steels from a number of commercial sources.

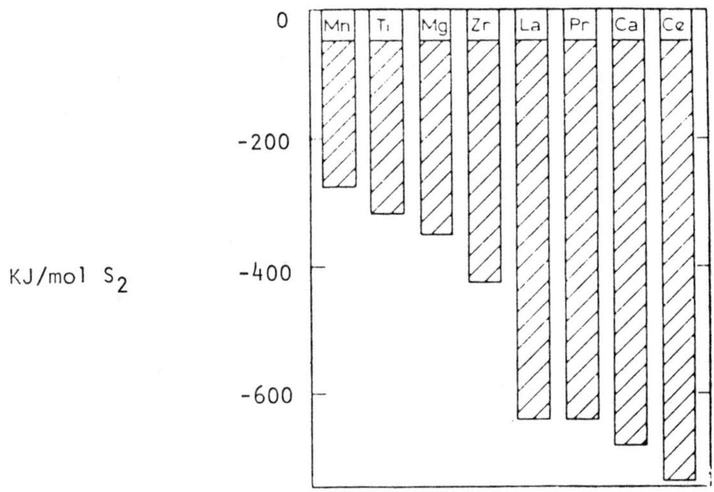

Fig. 1. Comparison of Standard Free Energies of Formation of Rare Earth and Other Sulphides at 1650° C(2).

The rare earth treated steel was also evaluated using "dynamic tear (DT) type" specimens with both ground notch & fatigue-precrack notch geometries, and by ball test (a production quality control test). Before using the steel for harrow disk production on a commercial scale, field tests were conducted to compare performance with disks made from cross rolled steel. The rare earth treated 1085 steel for disk production was somewhat different from our initial experimental rare earth treated steel, see Table 4.

Table 1. Melting Points of Rare Earth Compounds and Other Common Compounds Occurring in Steel(3).

COMPOUND	MELTING POINT, °C
MnS	1530
CeS	2099
Ce_2S_3	2149
LaS	1971
La_2S_3	2099
NdS	2138
Nd_2S_3	2199
Ce_2O_2S	1949
La_2O_2S	1993
Nd_2O_2S	1988
CeO	2599
Ce_2O_3	1691
La_2O_3	2249
Nd_2O_3	2271
Al_2O_3	2030
CaO	2600

Table 2. "Hypercal" Alloy*

ELEMENTS	WT. PERCENT
Ca	10-13
Ba	9-12
Si	38-40
Al	19-21
Fe	Balance

*Developed by Union Carbide Corp.

Table 3. Typical Rare Earth Alloys(2)

RE ALLOY	TYPICAL COMPOSITION, WT. PERCENT					
	TOTAL RE	Ce	La	Nd	Fe	Si
Mischmetal	96.0 min.	45-51	23-26	15-19	3.0 max.	----
RE Silicide	30-35	48-50	32-34	13-14	30-35	30-35

EXPERIMENTAL PROCEDURE

Test Materials*

The following materials were evaluated. Their chemical analyses are given in Table 4.

1. Commercial grade AISI 1085 steel.
2. RE-silicide treated AISI 1085 (50% globular sulphide) - experimental.
3. RE (mischmetal) treated AISI 1085 (100% globular sulphide) - experimental.
4. RE treated AISI 1085 (100% globular sulphide) - production disk.
5. Disk Source A: cross-rolled 1085.
6. Disk Source B: high Mn, low S 1085.
7. Disk Source C: special practice 1085.

*All the test materials were not evaluated by all the tests described in this paper.

Figs. 2-7 show the nature of the sulphides in conventionally rolled (straight rolled) steel, rare earth treated steel and cross-rolled steel.

Table 4. Chemical Analyses, Wt. Percent

SAMPLE	C	Mn	Si	P	S	Al	Ce
1085 Steel	0.84	0.86	0.19	0.015	0.025	0.047	0
RE-silicide treated	0.88	0.80	0.21	0.020	0.012	0.031	0.002
RE treated (exp.)	0.84	0.78	0.18	0.015	0.010	0.048	0.013
RE treated (prod.)	0.91	0.96	0.20	0.020	0.027	0.03	0.03
Disk Source A (cross-rolled)	0.84	1.00	0.25	0.020	0.020	0.04	0
Disk Source B	0.87	1.20	0.22	0.020	0.015	0.05	0
Disk Source C	0.91	0.93	0.17	0.020	0.022	0.017	0

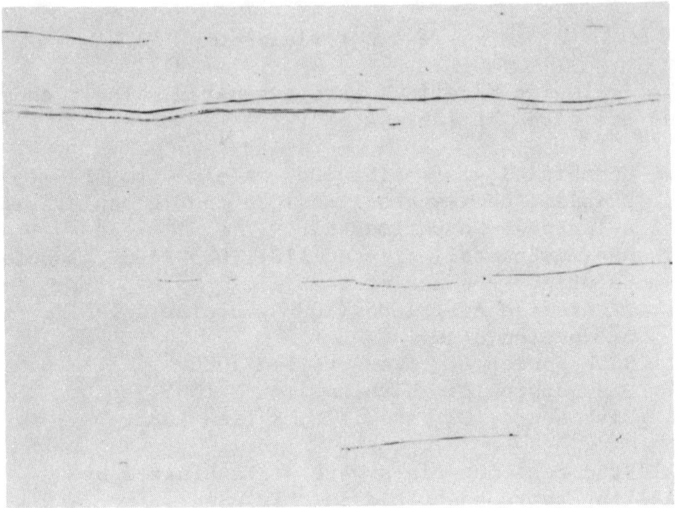

Rolling

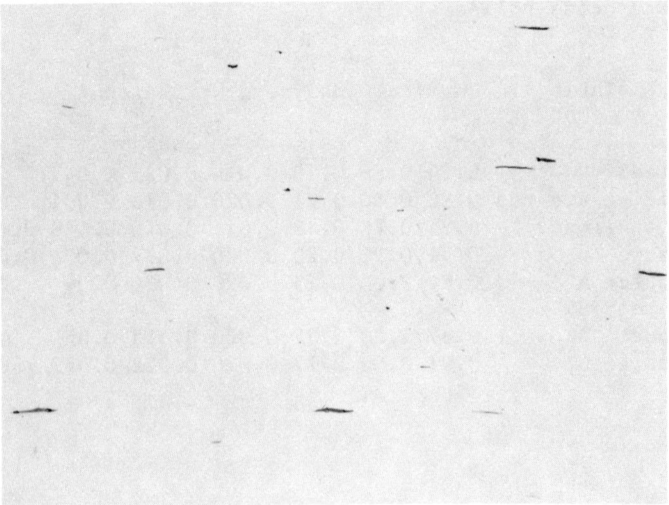

Unetched Transverse 500X

Figs. 2 & 3. The two figures show inclusions in the rolling direction & transverse direction respectively in straight rolled steel - commercial grade 1085, Source B, Source C.

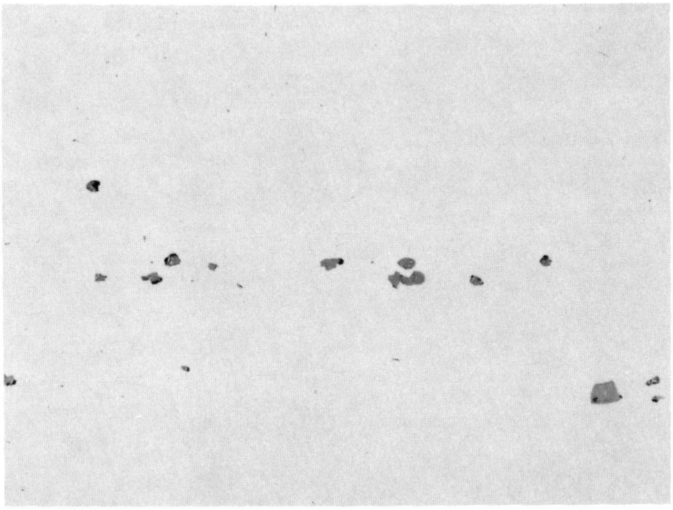

Rolling

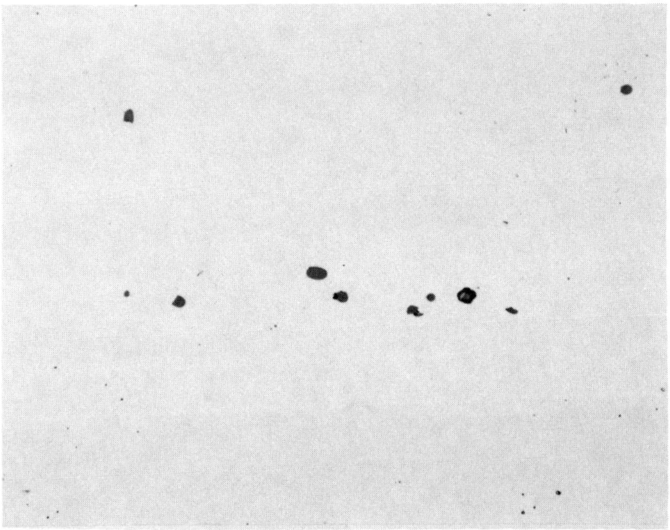

Unetched Transverse 500X

Figs. 4 & 5. The two figures show inclusions in the rolling & transverse direction respectively in mischmetal treated steel.

Unetched 500X

Figs. 6 & 7. The two figures show inclusions in the two rolling directions of cross-rolled steel - Source A.

Steelmaking

In order to change the morphology of sulphides in disk steel from stringer to spherical shape, additives such as titanium, calcium (Hypercal), rare earth silicide, and mischmetal were used. All these experimental steels were produced at Wisconsin Steel by treating AISI 1085 steel with the various additives. A full heat of 1085 steel was treated with "Hypercal" in the ladle. Other additives were placed into the ingot molds thus producing only one ingot for each treatment. The experimental ingots were rolled to 9" wide x 5/16" thick bars.

Titanium, "Hypercal" and rare earth silicide treatments were only partially successful in producing non-deformable globular sulphides whereas the mischmetal treatment produced 100% globular sulphides. Steelmaking practice concerning the mischmetal treated steel was as follows:

1. Steel was made in a basic oxygen furnace, BOF, 140 ton heat.
2. Ladle analysis: C 0.82, Mn 0.76, Si 0.22, P 0.024, S 0.016, Ni 0.03, Cr 0.03, Mo 0.01, Al 0.032%.
3. The first ingot (5 tons) was treated with mischmetal. Nine pounds of mischmetal were added by throwing 2 ounce balls into the ingot mold as it was filled. Addition was started after the bottom of the ingot mold was covered with molten steel and was continued in a steady manner until the mold was 3/4 full.

Rare earth treated 1085 steel for 20 inch diameter 7 gauge disk production was produced by a different steel company. This steel was somewhat different from our initial experimental RE treated steel, produced by Wisconsin Steel, with respect to chemical composition (see Table 4), and therefore, it has been identified as RE treated 1085 steel (production).

Heat Treating

5/16" thick bars were ground to 0.180 inch thickness (equivalent to 7 gauge disk) before heat treating. Test samples were austempered (bainite microstructure), and quench and tempered (martensite microstructure).

Austemper - Austempering procedure consisted of austenitizing at 1600° F for 4 minutes in a salt bath, quenching into a water saturated salt bath at 550° F for 1 minute, then quenching into another salt bath at 730° F for 20 minutes (hardness RC 40-44).

Quench and Temper (QT) - Quench and tempering procedure consisted of austenitizing at 1500° F, quenching in oil at 110° F, and tempering at 935° F (hardness RC 40-44).

Test Procedure

Fracture Toughness Test - Tapered double cantilever beam (TDCB) specimens, Fig. 8, were used to evaluate fracture toughness(4). The test specimens were fatigue precracked in stages by decreasing the load and using a stress ratio, σ min./σ max., of 0.1. A typical precracking procedure is: first stage 1200 pounds 128000 cycles, second stage 1000 pounds 76000 cycles, third stage 800 pounds 108000 cycles. The final fatigue crack was at least 0.05 inch in length.

Tests were conducted at room temperature at a rate such that the load increased from zero to the critical value (fracture load) within two minutes. Load-displacement plots were obtained with the help of an X-Y recorder and a clip type compliance gauge to measure displacement. The critical load (P_c) was determined in accordance with the ASTM procedure for fracture toughness testing(5).

Fracture toughness, K_c, values were calculated from the P_c values using the following equations(4):

$$G_c = \frac{4\,P_c^2}{E\,b_n\,b}\,m' \qquad K_c = \sqrt{G_c E}$$

G_c = Critical crack extension force (pounds/inch)
K_c = Fracture toughness (Ksi $\sqrt{\text{inch}}$), plane stress condition
P_c = Critical applied load (pounds)
E = Young's modulus (Ksi)
b_n = Crack width (inch)
b = Test specimen thickness (inch)
a = Crack length (inch)
h = Beam height at a distance "x" (inch), see Fig. 8
m = $\frac{3a^2}{h^3} + \frac{1}{h} = 3$
m' = Corrected value of m on the basis of the difference between the theoretical and experimentally determined values of m = 4.75.

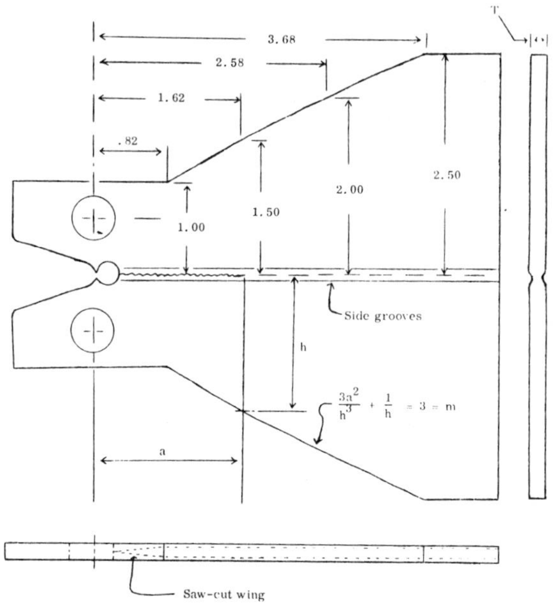

Fig. 8. Tapered-Double-Cantilever-Beam (TDCB) Specimen(4).

<u>3-Point Bend Test</u> - Dynamic tear type specimens, 5 inches long x 1 inch wide, were used in the 3-point bend tests, see Fig. 9. The thickness of the specimens was the same as 7 gauge disk thickness, 0.180 inch thick. Longitudinal and transverse specimens with ground notch (abrasive wheel cut 0.19 inch deep, 0.07 inch wide, notch tip radius 0.035 inch), and fatigue precrack (abrasive wheel cut 0.08 inch deep, jeweler's saw cut 0.01 inch deep, fatigue crack 0.10 inch deep) were tested. In most cases 2 longitudinal and 2 transverse specimens were tested for each material.

Tests were conducted with a computerized electrohydraulic material testing system (MTS) at a low deflection rate, loading ram travel 0.125 inches per minute. Load displacement charts and energy required to fracture were obtained with the aid of a PDP 11/35 computer.

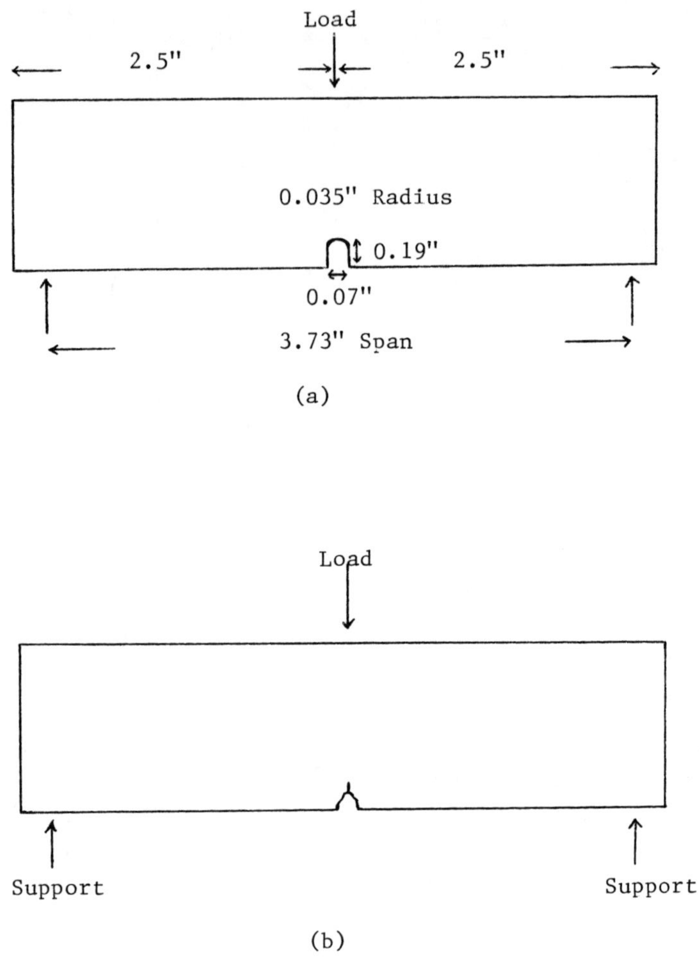

Fig. 9. 3-Point Bend Test Specimens (schematic sketch)
(a) Ground Notched
(b) Fatigue Pre-cracked: 0.08" ground notch, 0.01" jeweler's saw cut notch, 0.10" fatigue crack.

Ball Test - A 0.5 inch diameter hole is drilled through the sample at a minimum distance of 3 inches from any free surface. The sample is supported on a ring of 3 inches diameter and having 0.03 inch radius in the inner edge. A 1 inch diameter, hardened and polished ball is pushed into the hole. The test is considered complete when a crack propagates to the supporting ring or the ball passes through the hole.

The nature of the fracture and the maximum load required to push the ball are recorded. Short, ductile, non-directional fractures are desirable. Brittle fractures of any kind, and long directional fractures extending up to the 3 inch diameter supporting ring are undesirable.

Field Test - Field tests were conducted at Hagerstown, Maryland. The test area was extremely rocky, and is known to be conducive to disk fracture. The following 20 inch diameter 7 gauge (0.180 inch thick) harrow disks were used in the field test.

1. Rare earth treated (production) disks, austempered (bainite microstructure), Rockwell C 40-44 hardness.
2. Cross-rolled disks from Source A, quench and tempered (martensite microstructure), Rockwell C 40-44 hardness.

A 14 foot harrow was used with the two groups of disks alternatively positioned in the harrow.

RESULTS AND DISCUSSION

The chemical composition of the test materials are given in Table 4. All the test materials were essentially 1085 steel with minor variations in composition. Rare earth (mischmetal) treated steel contained 100% globular sulphide whereas the rare earth-silicide treated steel did not produce 100% morphology control because of an insufficient amount of RE as indicated by 0.002% Ce. However, it is believed that steelmaking conditions with respect to deoxidation and the amount of additives could be adjusted to improve the effectiveness of this treatment. RE treated steel for disk production contained higher C, Mn and S than our initial experimental steel.

Fig. 2 shows typical elongated sulphides in the rolling direction in commercial 1085 steel, in steel from disk Source

B and C. These sulphide stringers are a source of weakness in the rolling direction and contribute to directional behavior. Fig. 4 shows sulphides in straight rolled rare earth treated 1085 steel where the sulphides were found to be globular. Fig. 6 shows sulphide inclusions in cross-rolled 1085 steel from disk Source A where the MnS inclusions have a lenticular morphology. A cross-sectional view of these MnS inclusions shows short MnS inclusions in any direction.

The RE treated (experimental) 1085 steel was evaluated in both the austempered, and quench and tempered conditions in order to compare its performance with commercially available disks which are either austempered or quench and tempered. Fracture toughness (K_c, plane stress) data are listed in Table 5 and Fig. 10. As the steel was being developed for 7 gauge harrow disks, the samples were tested in 0.180 inch thickness which is equivalent to 7 gauge. This thickness was not sufficient to produce plane strain fracture toughness (K_{IC}) data. To obtain plane strain condition the thickness should be equal to or greater than $2.5 \left(\frac{K_{IC}}{YS}\right)^2$; YS = yield strength. In this case the thickness required to develop plane strain condition was estimated to be approximately 1.5 inch.

In these tests, both the austempered, and quench and tempered samples exhibited similar fracture toughness. The fracture toughness, K_c, of the RE treated steel was better than that of cross-rolled steel, and other steels. The isotropic characteristics of the RE treated steel is quite notable.

Table 5. Fracture Toughness, K_c

SAMPLE	HEAT TREATMENT	AVG. K_c, KSI $\sqrt{INCH}$ LONG.*	TRANS.**
Commercial 1085	Austemper	62	105
	Q and T	57	112
RE-Silicide Treated	Austemper	91	133
	Q and T	93	114
RE Treated (Exp.)	Austemper	119	121
	Q and T	115	120
Source A (Cross-Roll)	Austemper	99	112
	Q and T	97	110
Source B	Austemper	---	---
	Q and T	74	92
Source C	Austemper	82	86
	Q and T	78	114

*Long. - Specimen with crack propagation parallel to rolling direction.

**Trans. - Specimen with crack propagation perpendicular to rolling direction. For cross-rolled steel transverse direction represents another rolling direction.

Q and T - Quench and Temper

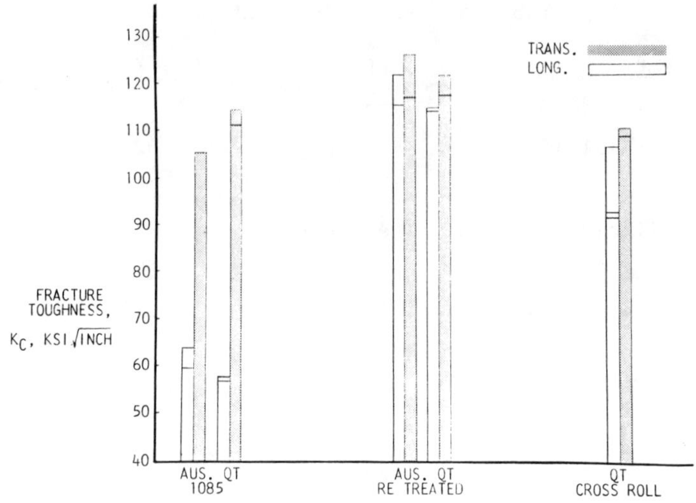

Fig. 10. Fracture Toughness Test Results. Each horizontal line on the bar chart represents one test. Aus. - austempered, QT - quench and tempered.

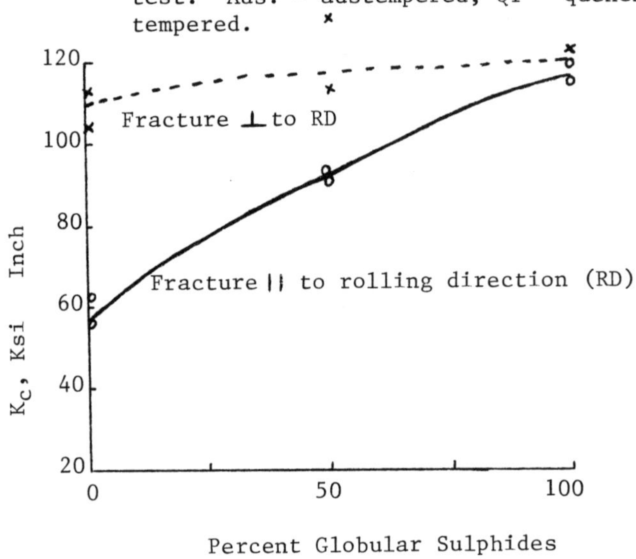

Fig. 11. Fracture Toughness vs Percent Globular Sulphides. Based on commercial 1085, RE -silicide treated and RE treated steel.

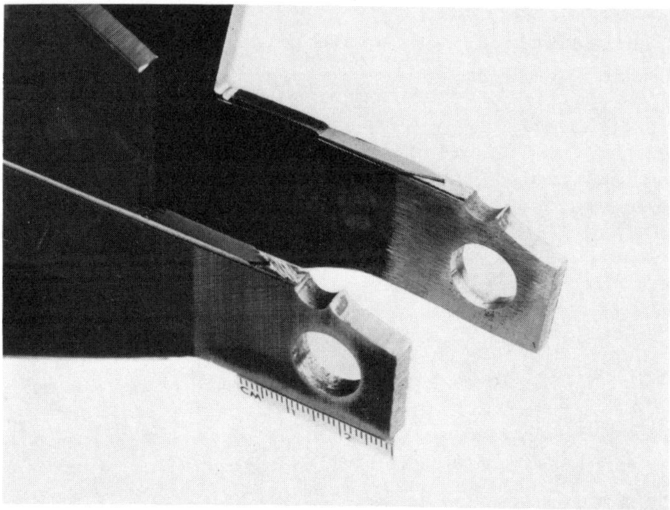

Fig. 12. Photograph shows tested fracture toughness specimens. In the left hand side specimen, 1085 steel, the crack propagated for a very short distance (beyond the fatigue precrack) along the groove. Then the specimen broke in the wing along the rolling direction in a brittle manner.

The right hand side specimen, rare earth treated 1085 steel, exhibits ductile fracture (beyond the fatigue precrack) indicating high resistance to crack propagation. Since the specimen started to bend, the crack could not be extended any more. A section from the wing has been cut to show the nature of the fracture.

Fig. 11 graphically portrays the effect of percent of globular sulphides on fracture toughness, K_c. These data are based on a limited number of tests, but there appears to be a trend that increased percentage of spherical sulphides increases fracture toughness in the longitudinal direction and thus significantly decreases anisotropy.

Fig. 12 shows typical fractures for commercial 1085 steel and the RE treated 1085 steel specimens. Commercial 1085 steel exhibited extremely poor resistance to fracture propagation. On the other hand the RE treated 1085 steel specimens fractured in a slant fracture ductile manner indicating very high resistance to fracture propagation. Specimens of the other groups exhibited mixed modes of fracture intermediate between these two extremes.

The microstructure of the austempered samples was upper bainite, and that of the quench and tempered samples was tempered martensite. Typical microstructures are shown in Figs. 13-14.

Fig. 13. Bainite Microstructure. RE treated (prod.) 1085 steel. Picral etch, 1000X.

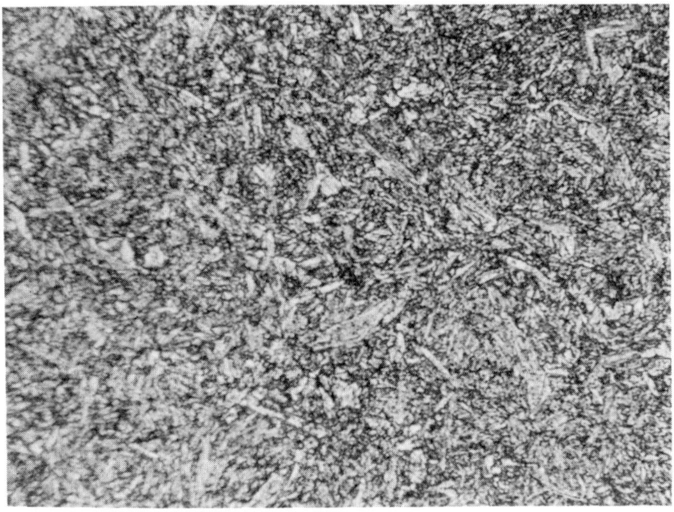

Fig. 14. Tempered Martensite Microstructure. Cross-rolled disk from Source A. Picral etch, 1000X.

The effect of hardness on fracture toughness, K_c, was investigated using TDCB specimens. In the hardness range of RC 40-50, the fracture toughness rapidly decreases with the increase in hardness, see Fig. 15.

The 3-point bend test results are given in Table 6 for the ground notched and fatigue-precracked specimens. Ground notched RE treated (experimental) 1085 and RE treated (production) 1085 steel in the austempered condition required significantly more energy to fracture than quench and tempered samples from disk Source A and B, and quench and tempered RE treated 1085. However, for the fatigue precracked condition, in general, austempered samples required lower energy to fracture than quench and tempered samples. The austempered RE treated (experimental) 1085 steel was an exception with wide scatter in data, 525 inch-pounds to 69 inch-pounds. The reason for this discrepancy has not been determined.

On the basis of overall performance the RE treated (exp.) 1085 was superior to all other samples including RE treated (prod.) 1085. Higher C, higher S and higher hardness (RC 42 vs RC 40) are some of the contributing factors for the difference in performance between the two groups of RE treated steel.

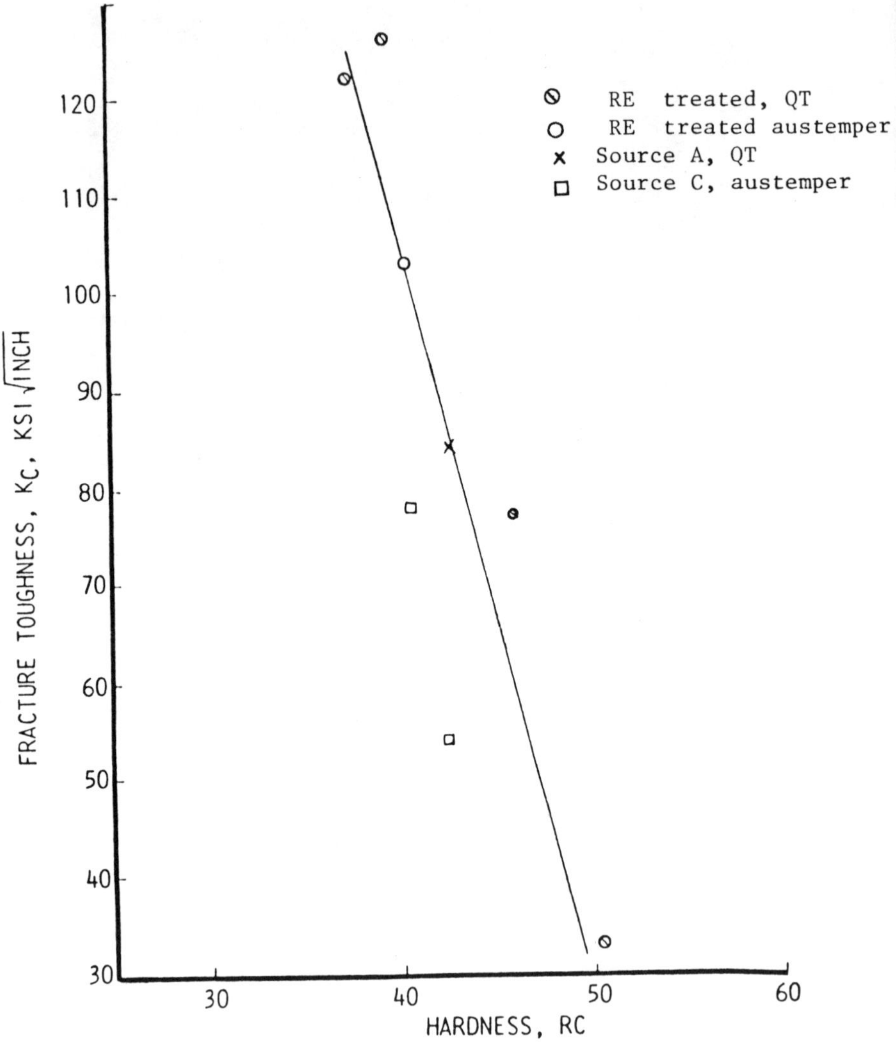

Fig. 15. Effect of Hardness on Fracture Toughness, Longitudinal TDCB Specimens.

Table 6. 3-Point Bend Test Results

SAMPLE	AVG. ENERGY TO FRACTURE, INCH-POUNDS			
	GROUND NOTCH		FATIGUE-PRECRACK	
	Long.	Trans.	Long.	Trans.
RE Treated 1085 (exp.)				
Austemper	794	840	69*	642
Q and T	360	400	89	241
RE Treated 1085 (prod.)**				
Austemper	568	673	38	34
Q and T	248	250	74	84
Disk Source A (cross-roll)***				
Q and T	227	303	80	142
Disk Source B				
Q and T	260	---	52	125

Long. - Crack propagation parallel to rolling direction.
Trans. - Crack propagation perpendicular to rolling direction.
*There was a wide range in the test data. This is the minimum value. Other values were 525 and 435.
**Production steel which was field tested with the cross-rolled disks from Source A.
***In this case transverse direction represents another rolling direction.

Fig. 16. SEM fractograph shows stringer type inclusions in the rolling direction. Disk Source B, 3-point bend specimen fracture, 100X.

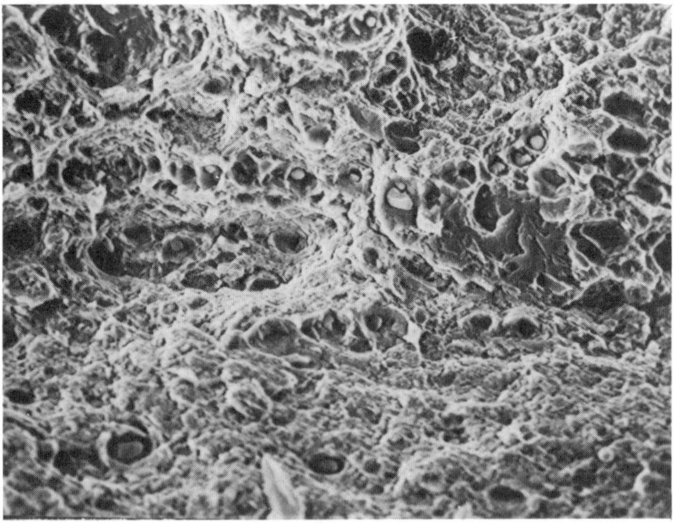

Fig. 17. SEM fractograph shows globular inclusions in the rolling direction. RE treated (prod.) 1085, 3-point bend specimen fracture, 1000X.

Fig. 18. SEM fractograph shows shorter inclusions relative to disk Source B. Cross-rolled steel from Source A in one rolling direction, 3-point bend specimen fracture, 100X.

Figs. 16-18 show the electron fractographs of 3-point bend specimens illustrating the nature of fractures in straight rolled steel, RE treated straight rolled steel and cross-rolled steel.

Ball test results are illustrated in Figs. 19-21. Samples from disk Sources B and C developed straight line directional fractures in the ball test. RE treated steel developed short, ductile and non-directional fracture. The ball test performance of the cross-rolled steel was comparable to that of the RE treated steel, however, austempered samples performed better than quench and tempered samples in the ball test with regard to the maximum load which can be obtained.

A field test was conducted at Hagerstown, Maryland, to compare RE treated (prod.) austempered disks with cross-rolled quench and tempered disks from Source A. In this test which was deliberately conducted in rocky soil known to be conducive to disk fracture, the RE treated disks performed better than the cross-rolled disks with regard to the amount of tearing occurring at the disk edges. Neither disk type sustained any straight line extensive cracking, see Figs. 22-23.

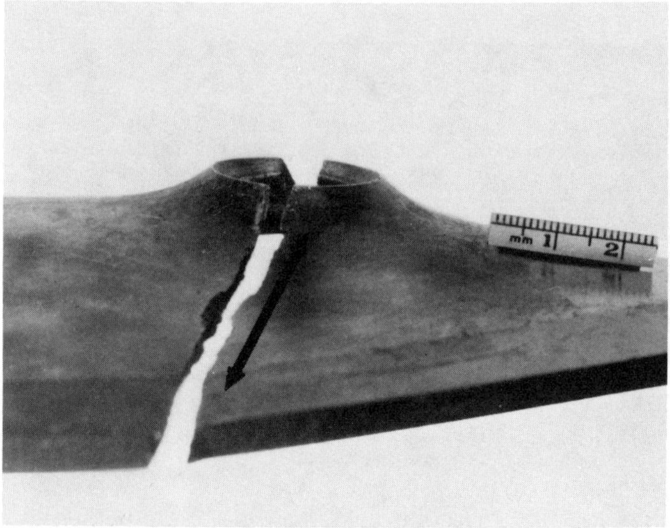

Fig. 19. Ball tested straight rolled disk flat, austempered, Source C, developed extensive crack in the rolling direction. Maximum load 27000 pounds.

Fig. 20. Ball tested RE treated (prod.) disk flat, austempered. Short, ductile, non-directional fractures. Maximum load 25000 pounds.

Fig. 21. Ball tested cross-rolled disk flat, quench and tempered, Source A. Short, ductile, non-directional fractures. Maximum load 19000 pounds.

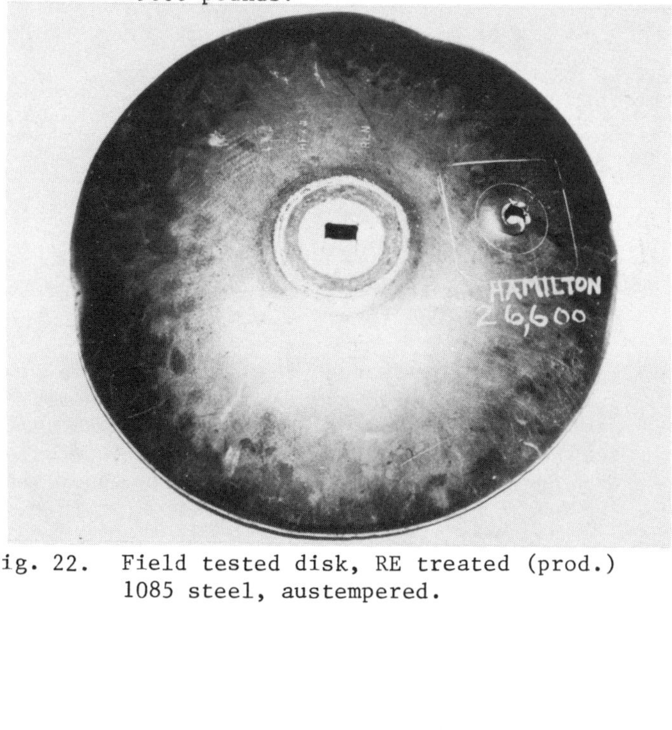

Fig. 22. Field tested disk, RE treated (prod.) 1085 steel, austempered.

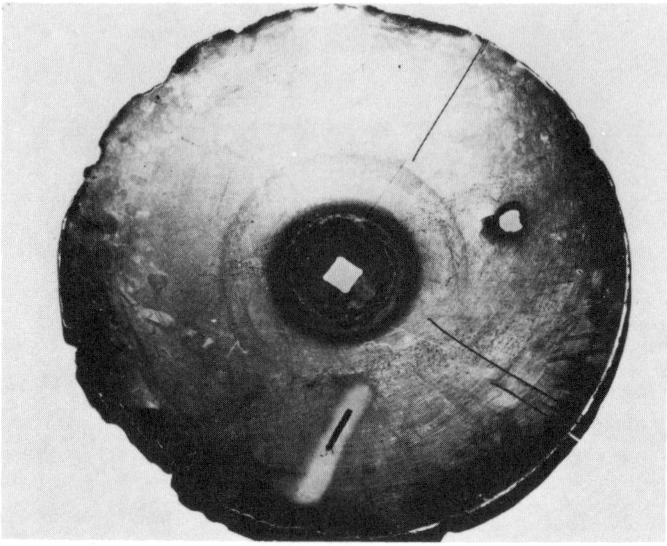

Fig. 23. Field tested disk, cross-rolled 1085 steel, quench and tempered, Source A.

SUMMARY

1. Rare earth treated 1085 steel is equivalent to or better than cross-rolled 1085 steel with respect to fracture toughness isotropy.

2. 3-point bend test:

   (a). In the ground notch test condition, austempered (bainite) samples performed better than quench and tempered (martensite) samples.

   (b). In the fatigue-precrack test condition, in general, quench and tempered samples performed better than austempered samples. The austempered RE treated (exp.) 1085 steel was an exception with wide scatter in data - very high value (525 inch-pounds) and low value (69 inch-pounds). The reason for this discrepancy has not been determined.

   (c). Increase in carbon, sulphur and hardness adversely affected notch toughness.

3. The ground notch test condition correlated with ball test and field test results of harrow disks, whereas the fatigue precracked test condition did not show a correlation.

4. Rare earth treated 1085 steel has been successfully applied for the production of heavy duty, superior quality harrow disks (U.S. Patent No. 4098 622) eliminating the need for cross-rolled steel.

REFERENCES

(1) L. Luyckx et al., "Sulphide Shape Control in High Strength Low Alloy Steels", Met. Trans., Vol. 1. 1970, p. 3341.

(2) L. Luyckx and J. R. Jackman, AIME Electric Furnace Conf., Vol. 31, 1973, p. 175.

(3) C. J. Bingel and L. V. Scott, AIME Electric Furnace Conf., Vol. 31, 1973, p. 171.

(4) S. Mostovoy, P. B. Crosley and E. J. Ripling, "Use of Crack-Line-Loaded Specimens for Measuring Plane-Strain Fracture Toughness", Journal of Materials, Vol. 2, No. 3, 1967.

(5) ASTM - Proposed recommended practice for plane-strain fracture toughness testing of high strength metallic materials using a fatigue-cracked bend specimen. ASTM Standards, Vol. 31, 1968.

NODULAR IRON CONVERSIONS ON A 4-WHEEL DRIVE AGRICULTURAL TRACTOR

Bob L. Wilson
John Deere Foundry

When four-wheel drive tractors were introduced in our Company, there was little use made of nodular iron on the early models. A number of factors contributed to this decision. Planned low volume due to uncertainty of customer acceptance, foundry facilities inadequate to produce castings of the size and complexity required, and the lack of interaction between the foundry, manufacturing engineering, and product engineering early in the design stage. This resulted in approximately 600# (272 kg) of nodular iron per tractor being used.

Early in the planning phase of model updating material utilization became a major objective. Engineering closely studied the various materials that would appear to give them the best value considering cost, function, and styling. Manufacturing Engineering and the Foundry Casting Engineering Departments were brought into the early feasibility studies, and became deeply involved in the material selection evaluations.

In order to determine which components and especially fabrications would give the best increased value to the corporation as a whole, fabrications on the "then" current model tractor were studied.

In determining which parts would be further evaluated for possible change, a feasibility study was conducted.

Close attention was given to castability, ease of manufacturing, and cost. The parts which contributed most to improved corporate profits were selected. This improved profit came about in five major areas:

A. Improved Function
B. Lower Over-All Cost
C. Improved Manufacturing
D. Increased Tonnage Poured in the Foundry
E. Improved Appearance

This discussion will be concerned with seven parts forming a "Family" of heavy duty components. When assembled together, they form the backbone of the tractor carrying high loads at concentrated points. They are:

A. The Front Frame (Fig. 1)
B. Front Axle Support (Fig. 1)
C. Front Hinge (Fig. 2)
D. Right Cab Support (Fig. 2)
E. Left Cab Support (Fig. 2)
F. Rear Hinge (Fig. 3)
G. Steering Cylinder Bracket (Fig. 3)

The following design goals were set:

A. The Hinges must perform several functions.
    1. They retain power train components
    2. They are part of the chasis structure
    3. They must be strong to serve as the main power connecting link, and pivot mechanism between the front and rear part of the tractor.
    4. The front hinge must support the operator enclosure.
B. High strength and reliability are required.
C. Low number of parts to simplify manufacturing.
D. Economical to produce.
    1. Each converted part must deliver a greater value to the corporation over the part it replaced.

The first component we will discuss will be the FRONT HINGE CASTING. (Fig. 2) As mentioned earlier, our first design goal was for the hinge to retain power train components which the front hinge in its fabricated state could not do. (Fig. 4) An additional casting in the form of a "Hinge Cover" containing bearing support provisions would have to be provided. This cover would require the same machining in the bearing bore areas as the nodular cast hinge now required. Two additional interface surfaces for fit up would be necessary; one on the fabricated hinge and two on the cover casting. The fabrication also had two large steel plates welded to the hinge. These plates served as members carrying the cab enclosure. Due to the loads these .62 (16mm) thick plates carried, it was necessary they be welded to the top of the hinge in an upright position (Fig. 4). This design satisfactorily performed its function, however, due to its relatively light mass and the "tuning fork" shape they formed, the hinge transmitted noise into the cab enclosure area.

In the casting we were able to accomplish the goal of supplying provisions for the power train components (Fig. 2). The casting also provides a better way of attaching the soundgard cab supports. The bolted joint plus the additional mass in the casting reduced noise transmitted into the cab enclosure (Fig. 2). The hinges are a part of the chasis structure, a function that both the fabrication and casting performs.

The present fabrication was interpreted as a casting by the Foundry and design recommendations were made. These recommendations resulted in reducing the number of cores required and an increase in percentage of yield by adding and removing material to obtain a more uniform section thickness where it was possible. This is important to the Foundry in any cast shape, but especially so in complicated castings such as the hinges under consideration.

Using this information, the first experimental casting drawing was completed. The first build on the new model tractors were started using basically .75 (19mm) wall sections throughout the hinge castings (Fig. 2). These first cast hinges were placed in field test tractors to be run under field conditions. In conjunction with service hours being accumulated in the field, a stress tractor was also built. Stress tests were conducted in the laboratory using brittle laquor and strain gauge. Midway between first and second build of experimental tractors, information gleaned from field test, and stress test tractors were used to conduct a finite element study of the hinge casting design.

The finite element study indicated we were over designed. This was confirmed by further stress testing. The basic section size was reduced from .75 (20mm) to .51 (13mm) walls on the second build tractors (Fig. 2). Ribs were added to strengthen areas of high stress.

REAR HINGE, A single casting (Fig. 3) replaced 54 fabricated parts. The casting design now allowed the drive line bearings to be run in oil. We were able to go from a dry clutch to a wet clutch, again without the use of an extra cover casting. This extra casting would require about the same machining to adapt to the fabricated hinge as is now performed on the cast nodular rear hinge. Finite element study showed a need to also reduce section sizes on the Rear Hinge, and a .51 (13mm) wall was adopted for testing. The casting provided better steering cylinder attaching points (Fig. 3 and Fig. 4). However, in this area our finite element study showed the cross sectional area should be increased from .79 (20mm) to 1.06 (27mm). The casting also allowed us to incorporate the power take-off brake into the rear hinge. This was accomplished by casting material and a groove 4.97ID x 6.97OD x .87 (126mmID x 177mmOD x 22mm) deep around the bearing bore carrying the P.T.O. Shaft (Fig. 3). The brake is hydraulic and operates off oil used for the power train compartment. Inner connecting passage ways were cast into the oil side of the hinge (Fig. 3). The cast passages connect the power train bearings which are not submerged in oil. These features were supplied at little or no extra cost in the casting, and when compared with other means of providing them, substantial savings were realized.

In areas of high stress we were able to add material easily which was difficult to do in steel. Fit up of parts was much improved due to inconsistences in the fabrication caused by stresses when welding this large number of pieces together. Improved appearance added to the overall styling of the tractor.

The rear hinge casting design was run concurrently with the Front Hinge and the same procedures for design study were used. Due to the success of the two hinge conversions on the first build tractors, it was decided to proceed with the other parts for second build tractors.

The FRONT AXLE SUPPORT (Fig. 1) provides attaching points for the engine and main frame of the front half of the tractor to the differential and planetary front axle assembly. The cast support replaced a fabricated support which had 16 parts welded in the assembly (Fig.4).

A casting design was generated using input from finite element and information gained from strain gauges on the first build stress tractor. This information showed .51 (13mm) section thickness should be used. The only problems encountered were in the Foundry. This section thickness spread over this large flat surface, 29 x 33 (738mm x 831mm), was difficult to pour without encountering misrun or cold-shut (Fig. 1). This problem was overcome through gating practices, some localized design modifications where these defects occurred, and controlling pouring rate and temperature. The results were a more accurate support, more economical to produce, improved manufacturing, and easier final assembly.

CAB SUPPORTS, (Fig. 2), Right and left hand cab support castings are used. These are mounted to the front hinge and support the cab at four attaching points mounted on a sound absorbing material. Because of the safety related application of these parts, a very high level of confidence in design and casting quality were required of the castings. Rigorous test procedures were conducted on the support structure. The castings were tested by a roll-over of the tractor with the cab super structure mounted while traveling in a forward direction. An impact test using a steel ball swung as a pendulum in a 0° fahrenheit environment caused no problems.

STEERING CYLINDER BRACKET, (Fig. 3) This casting replaced a steel fabrication containing ten common parts welded together (Fig. 5). The bracket serves as an anchor point for the steering cylinders. The casting proved to supply a more rigid anchor point and was lower in cost.

The FRONT FRAME (Fig. 5) consisted of a formed and punched member with two bosses welded for attaching to the tractor. The frame provides receptacle and carries the front lights. The part connects the side frames together forming the lower front styling of the tractor. Because the part is located in a position for all to see it must present a pleasing appearance.

In its fabricated form, the front frame performed its mechanical function as well as the casting. However, as improved appearance was a design goal on the new models, this proved difficult to attain with the fabrication. A number of design goals were satisfied by using a casting.

A. It functions as a rigid connection for the side frames.
B. It provides protection for and serves as a carrier for the front lights at little or no extra cost.
C. By the ability to place metal in the desired shape economically, the casting contributes to the overall design lines of the tractor.
D. By using a casting in place of a fabrication on the new model tractors, a 25% cost savings was realized.

This discussion has dealt with seven parts and does not represent all parts that were converted to castings.

As mentioned earlier, the first production model tractor contained approximately 600# (272 kg) of nodular iron. The new updated models were adopted with 3,100# (1,406 kg) of nodular iron used per tractor. This is a good example of what can be accomplished in the material selection process where various manufacturing and engineering disciplines are brought into a project in the early stages.

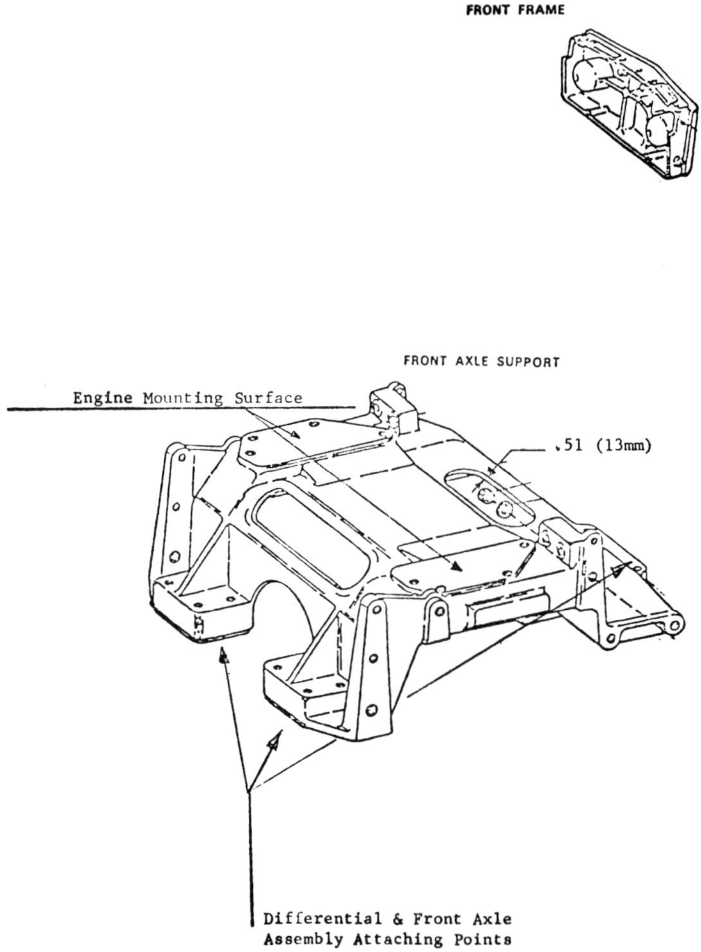
FRONT FRAME
FRONT AXLE SUPPORT
Engine Mounting Surface
.51 (13mm)
Differential & Front Axle
Assembly Attaching Points

FIGURE 1

SOUND-GARD BODY MOUNTING

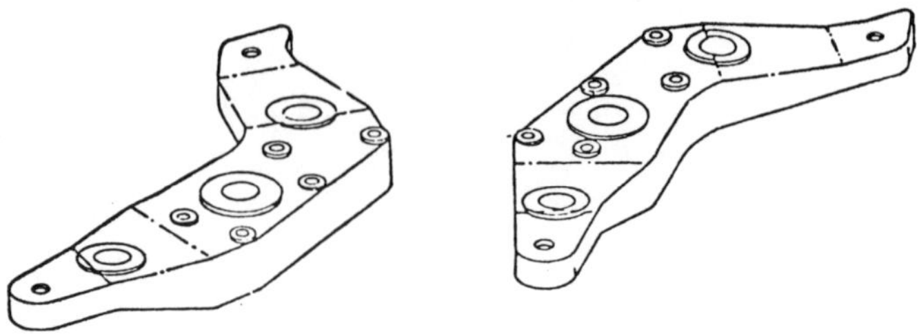

FRONT HINGE

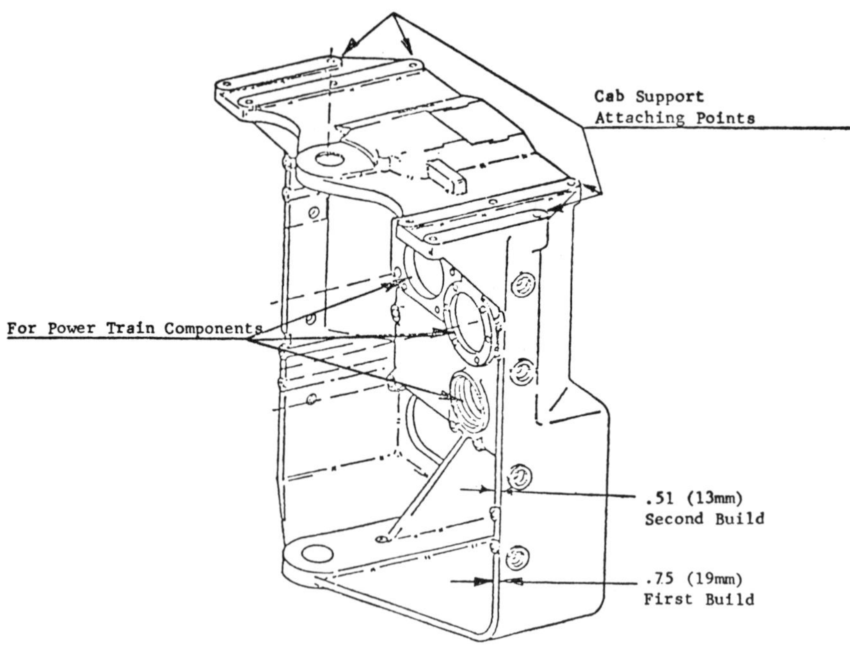

FIGURE 2

STEERING CYLINDER BRACKET

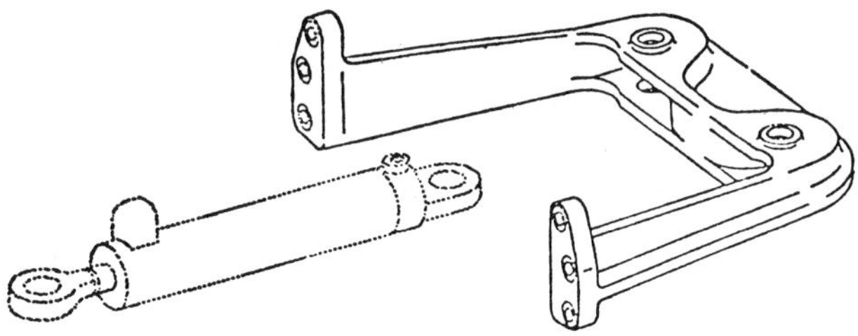

REAR HINGE

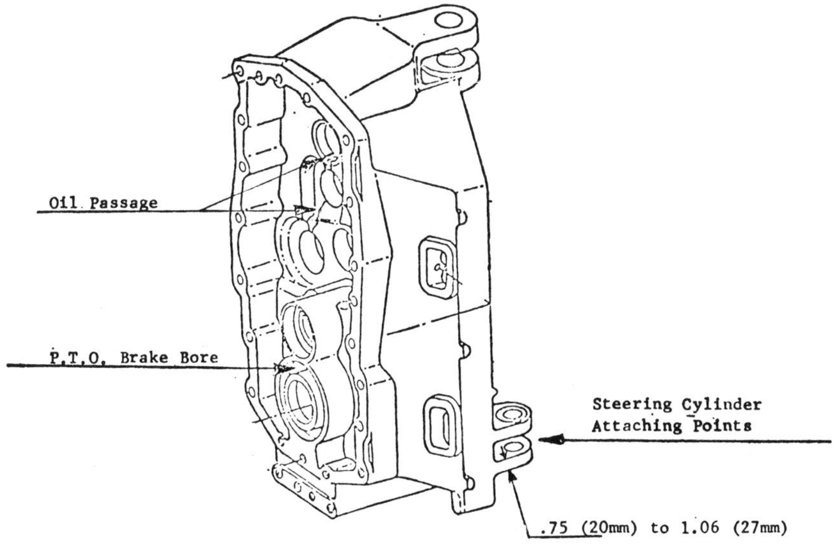

FIGURE 3

FRAME

FRONT AND REAR HINGES

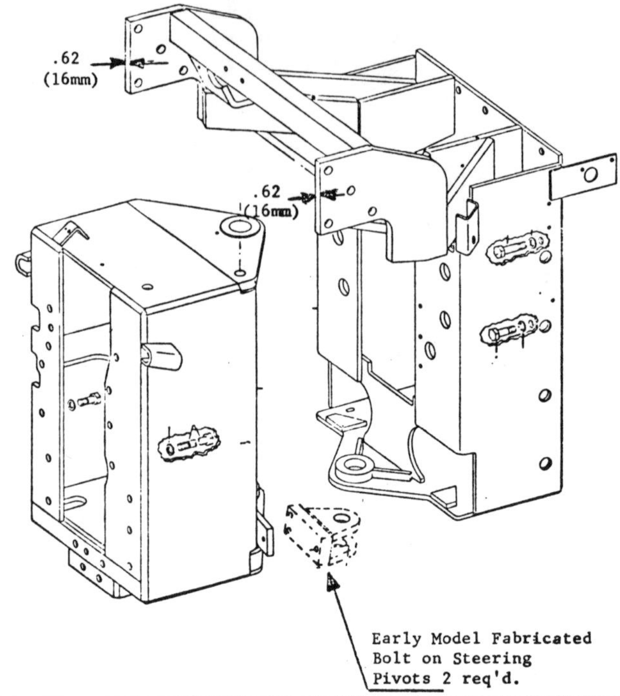

FRONT AXLE SUPPORT

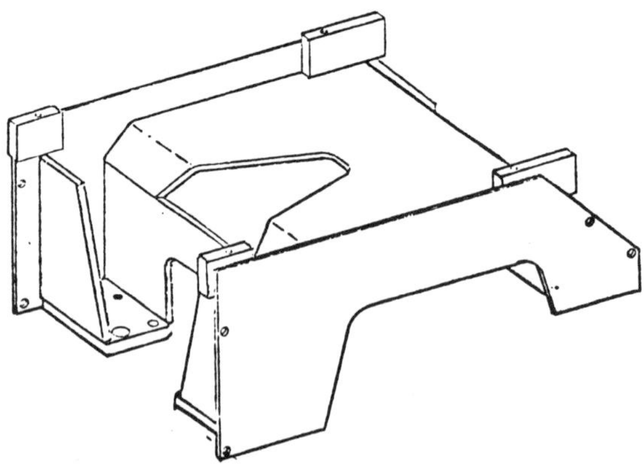

FIGURE 4

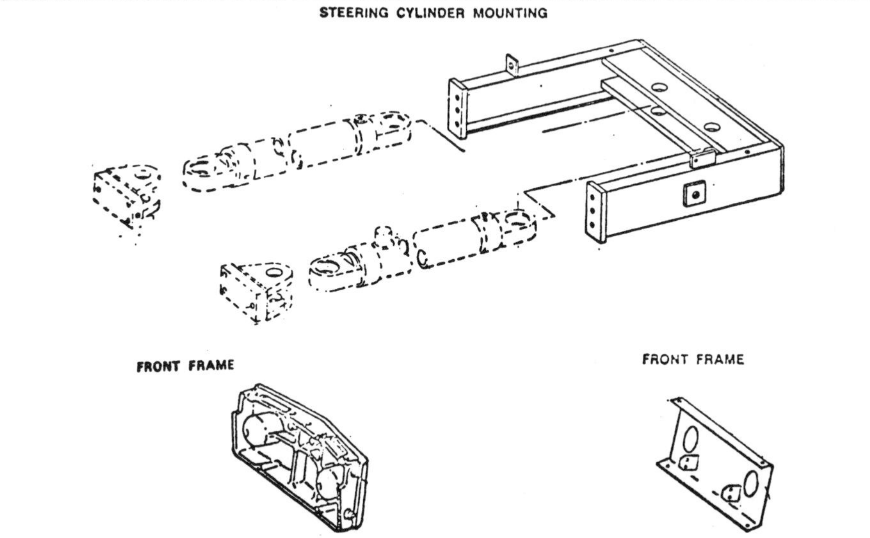

FIGURE 5

DESIGN EFFICIENCY — II:

ADHESIVES AND SEALANTS IN THE STRUCTURAL WORLD

ADVANCES IN ANAEROBIC ADHESIVE TECHNOLOGY

BRUCE D. MURRAY
LOCTITE CORPORATION
705 North Mountain Road
Newington, Connecticut 06111

I. INTRODUCTION

Anaerobic adhesives and sealants are liquid materials which are stable when stored in contact with air. The oxygen in the air prevents the chemical reaction which causes cure, from proceeding. Placing anaerobic adhesives between metal parts out of contact with air accelerates the cure reaction, and the anaerobic adhesives and sealants will cure at a predetermined rate. Cure rate can be controlled to allow very rapid assembly of parts in high speed production systems, and also assembly and adjustment of parts with slow curing system.

Thus, the combination of metal contact and the absence of air will cause anaerobic adhesives to cure readily to a hard, tough, cross-linked thermoset plastic. The anaerobic adhesives and sealants exhibit only minimal shrinkage during cure and thus they are effective in many areas for holding parts together and sealing assemblies against leakage and internal corrosion. The cross-linked polymers are effective in environments with temperatures up to 450°F and will not readily degrade in organic solvents or mild acids or bases.

The nature of anaerobic formulation has allowed extensive modification of properties such as cure speed, viscosity, cured polymer hardness, adhesive strength and environmental resistance. This property modification capability has resulted in a wide variety of adhesives and sealants for a number of use areas. The unique single component cure system has resulted in significant benefits for users. Basically, these benefits translate into cost savings on a production line resulting from weight savings, reduction of operations, and elimination of assembly problems on the line. Parts assembled with anaerobic adhesives also have increased reliability due to higher strength, improved vibration resistance and an increased resistance to environmental stresses. Anaerobic resin systems are also safe to use in that liquids are classified as mild skin sensitizers and generally will not harm skin of users. Liquids also have mild odors and are non-flammable.

Anaerobic technology basically fits into four specific use areas. These are locking threaded assemblies, sealing threads, flanges, or porosity in metals, retaining smooth cylindrical parts and structural bonding of flat metal, glass or plastic parts. Specific anaerobic adhesive and sealant properties have been developed for each of these use areas. Important aspects of anaerobic technology in each use area are discussed below. Significant new properties have been developed on the basis of new user requirements, which may be unfulfilled by available products. A comparison of anaerobic adhesive property developments to date is discussed in Table I.

Table 1. Property Development Anaerobic Adhesives

PROPERTY		1953	1979
Viscosity		15 cps	4-1 Million +
Full Cure Time	Inactive Metals	No Cure	6-24 hrs.
	Steel	24-72 hrs.	1-72 hrs.
Thixotropic Ratio		None	10-1
Tensile Strength		1,400 psi	8,000 psi
Shear Strength		1,000-2,000 psi	75-8,000 psi
Temperature Limits		-65°F to 300°F -55°C to 150°C	-65°F to 450°F -55°C to 230°C
Impact Strength		1 ft./lb.	1-50 ft./lbs.
Peel Strength		1 piw	1-40 piw

II. THREADLOCKING

The unique nature of anaerobic adhesives made them ideal for locking threaded assemblies. Early anaerobic products had little adhesive capability and thus were not suitable for structural adhesive bonding. However, the capability to control polymer hardness and achieve cure on metal surfaces allowed formulations which would fill all gaps in metal fasteners as liquids harden to a solid, tough polymer without shrinkage, and then "jam" the threads against movement, preventing loosening. Control of the polymer hardness has given us formulations which will not only prevent vibration loosening but also allow easy disassembly when required. Other formulation variations contain properties which match cure speed to requirements, give rapid reliable cures on oily "as received" fasteners, reduce scatter of fastener clamping forces by the incorporation of lubricants, and eliminate migration of uncured anaerobic adhesives away from the joint.

Anaerobic adhesives in threadlocking applications result in significant benefits to users. Weight is reduced by the elimination of metal locking devices with an insignificant weight of anaerobic resin. The locking capabilities of anaerobic adhesives is unsurpassed by other methods (1). Improved technology has allowed another step toward "the ideal locking device" with preapplied solid anaerobic locking compounds which will cure only after assembly of the threaded fasteners. The preapplied anaerobic is a dry to the touch coating not easily damaged by handling and will remain stable for extended periods of time. Curing and cured properties match those of liquid anaerobic properties. From the user point of view, these locking materials closely approximate the "ideal threadlocking device".

III. ANAEROBIC SEALING COMPOUNDS

Anaerobic formulations for sealing fit into 3 major categories. These are sealing threaded assemblies, an extension of threadlocking technology; sealing flanges in place of or in combination with solid gaskets; and sealing porosity in castings or powdered metal parts. The anaerobic cure mechanisms allow rapid cures without shrinkage so that gaps between metal parts are completely filled and sealed positively. Formulations on both ends of the viscosity spectrum serve for sealing applications. Very high viscosity formulations are used for sealing threaded assemblies against low pressure before curing and can be applied by the unique assembly method of silk screen in gasket applications. The silk screening technique allows a very precise placement of precise amounts of anaerobic resin, all within a short period of time under production line conditions.

Porosity sealing requires a very low viscosity material which flows readily into porosity in the metal. Anaerobic sealants can be applied to porosity through extensive vacuum impregnation processes required for filling large metal parts such as automotive engine blocks to methods as simple as brushing or spraying the resin onto the surface to seal weld porosity in telephone pole transformers. Excess low viscosity resin can be easily cleaned from the surface of the sealed parts and does not interfere with other finishing operations such as painting or plating.

IV. ANAEROBIC RETAINING COMPOUNDS

The assembly of smooth cylindrical parts has been traditionally accomplished by use of press fitting (machining the shaft larger than the hole into which it is assembled) or the use of mechanical fastening devices such as splines, keyways, or split rings. Advances in adhesion have resulted in anaerobic retaining compounds which will bond cylindrical pieces to strength as high as 6,000 psi. These "retaining compounds" can be used on shafts with rough surfaces and will cure through gaps of .015" or more. The use of retaining compounds has simplified small motor manufacture by mounting motor shafts into rotors and fans and eliminating requirements for shaft straightening operations. The loose fit allowed with retaining compounds can also eliminate pressing operations which often result in damaged parts. A high degree of adhesion and toughness of the cured anaerobic retaining compounds results in a "unitized" assembly which is shock resistant. The completely filled bond line also eliminates internal corrosion which is often found in press fitted or splined assemblies.

V. ANAEROBIC STRUCTURAL ADHESIVES

The newest anaerobic adhesive technology is the structural adhesive area which combines urethane technology with anaerobic acrylic technology. Anaerobic structural adhesives differ from other anaerobic formulations in that parts which they hold together are unsupported by mechanical fasteners. The cured anaerobic adhesive resin holds entire assemblies such as speaker magnets, automotive window trim, or other pieces through exposure to stresses and severe environments. Anaerobic structural adhesives exhibit strength properties as illustrated in Table 2. The true utility for users of anaerobic structural adhesives is in the area of manufacturing operations.

Table 2. Property Profile-Anaerobic Structural Adhesives

PROPERTY	VALUE
Tensile Shear - Steel	3,000-6,000 psi
- Oily Steel	1,500 psi
Tensile Strength	3,000-10,000 psi
Impact Strength	8-50 ft-lbs/in^2
Fatigue Strength	800-1,500 psi
T-Peel Strength	5-50 piw
Temperature Range	-80°F to 450°F -60°C to 230°C
Gap Curing and Bonding	to .080"
Elongation	50%+

Structural adhesive can eliminate many problems associated with mechanical fasteners such as loose fasteners, extra drilling and tapping operations, and excess weight associated with metal fasteners. In many cases, the use of adhesives has resulted in stronger, lighter and more durable assemblies. Adhesive use has also allowed use of new materials not easily assembled with mechanical fasteners.

Anaerobic structural adhesives form bonds as strong as most other kinds of adhesives. Anaerobic adhesive users find that this classification of structural adhesives results in a simpler on line assembly method and lower costs resulting from low energy cures and low labor automated assembly systems without sacrificing strength and durability.

Anaerobic structural adhesives are single component adhesives, which cure rapidly when used with surface activators.

They can also be cured using heat and, in some cases, ultraviolet light. Anaerobic structural adhesives are stable in automatic application systems, and can be applied through needle tips, by brush applicators, and in some cases, by silk screening. They form strong bonds to metals, glass, some types of wood and many types of plastics. Generally, an on line production bonding operation can take place within 5 minutes from the time of adhesive application to the time when parts are ready for other operations.

VI. ADVANCES IN ANAEROBIC TECHNOLOGY

Anaerobic technology improvements are based on new industry requirements. Improvements are generally intended to increase the scope of anaerobic adhesives and to improve on line reliability and assembly durability. Significant advances have been made recently in areas of temperature resistance, cure systems, and the capability of bonding a wider variety of materials reliably.

Table 3. Heat Ageing of Retaining Compound 620 on Steel*

---- "old" formulation
—— "new" hot strength compound

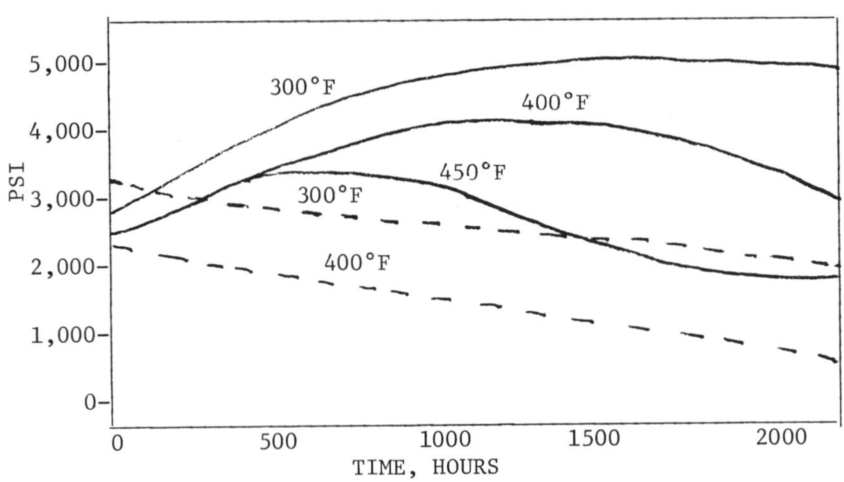

*Tested at temperature

Table 4. Ageing of Electron Beam Cured Adhesive at 232°C
Single Lap Shear Strength Mpa (psi)
Tested at 232°C After Ageing at 232°C

Adhesive	Initial at 20°C	70 hrs.	500 hrs.
Loctite 320-18	10.1 (1460)	8.8 (1280)	5.8 (840)
Loctite LO-559	9.9 (1430)	9.7 (1405)	9.2 (1340)
Hexcel F-178	3.1 (450)	3.3 (480)	3.0 (430)
Modified Kerimid 601	2.8 (410)	2.7 (385)	2.8 (410)
Modified Hexcel F-178	3.0 (430)	3.2 (460)	2.9 (420)

Generally, the upper limit for use temperature of anaerobics has been approximately 300°F. In combination with other resin systems, anaerobic resins have been successful in operating for short periods at temperatures as high as 400°F. New additive technology, however, has increased the temperature limit. Table 3 compares heat ageing of a new retaining compound and a previous high temperature retaining compound. The new resin system shows improved hot strength after initial exposure to temperature and far surpasses the previous material at temperatures as high as 450°F. Tests run on a developmental structural adhesive (Table 4) have indicated that this approach also shows excellent performance in other anaerobic adhesive systems, and these systems can surpass performance of other adhesive types.

In 1975 a series of anaerobic formulations were introduced with a much higher flexibility and greater toughness (2). These resins were used extensively in structural adhesive formulations. More recent work has led to improved retaining compounds and improved gasketing compounds which give a much higher degree of flexibility and offer a true structural capability to flanged assemblies. Retaining compounds formulated with these resins are tougher and more flexible, and thus are better able to successfully hold larger shaft hub assemblies, which are more subject to severe shock and stresses.

New cure systems have been also developed to eliminate some of the limitations of anaerobic adhesives. The newer "tough" anaerobic resins are suitable as coating materials and potting materials to protect metal surfaces and pot or encapsulate electronic equipment and components. Standard anaerobic chemistry would be suitable for tests such as these because of air inhibition. New cure systems which cure under high intensity ultraviolet light have been found to be acceptable for some coating and potting applications when light emitted from a mercury vapor lamp cures the entire resin system including the surface. This technology has been found particularly useful in the electronics industry for holding wires on electric motors and coils. Cures can be developed within 10 seconds in these areas. These new resins have been found suitable for UV light curing, heat curing and curing under electron beam radiation. Recent studies have indicated superior performance for anaerobic resins cured by electron beam (3).

The most dramatic and far reaching advance in anaerobic chemistry has been the capability to modify adhesive formulations with rubber compounds. Similar to acrylic adhesive technology, (which is sometimes based on anaerobic cure systems) the new rubber modified anaerobic develops many of the strength properties of acrylic adhesive technology, but maintains the rapid cure properties and low toxicity and flammability aspects of anaerobic adhesives. Table 5 illustrates properties of the first rubber modified anaerobic adhesive.

Table 5.

RUBBER MODIFIED ANAEROBIC ADHESIVE PROPERTIES	
Tensile Shear - Steel - Oily Steel - Aluminum	3,000 psi 1,800 psi 2,500 psi
Impact Strength	12 ft.-lbs.
T-Peel Strength	30 piw
Temperature Range	-60°F to 250°F
Gap Curing	.030"
Elongation	30%

Table 6

DEPEND™ PLASTIC BONDING PROPERTIES TENSILE SHEAR STRENGTH	
SUBSTRATE	STRENGTH, psi
ABS	560*
Acrylic	900*
Nylon	350
Phenolic	1,100
PVC	850*
Polystyrene	300*

* Substrate failed before bond.

Additional improvements inherent in the new anaerobic formulation are the capability of bonding the oily steel and capability of forming strong, durable bonds to thermoplastics such as ABS, acrylic, and polycarbonate. Plastic bonding properties are illustrated in Table 6. The combination of properties in these new resins enable potential anaerobic structural adhesive users to work with larger bonding surfaces, non-metallic surfaces and poorly prepared metal surfaces. Cures occur within minutes and bonds formed are strong, tenacious and resilient. Use areas for this adhesive extend to maintenance and repair of machinery operating in oily environments, and assembling larger components such as solar heat exchanger panels and railway car components.

VII. ANAEROBICS IN THE FUTURE

Anaerobic adhesive and sealant technology will continue to advance as new property requirements arise and as new technology is developed. The unique nature of anaerobic adhesives allow them to be used in combination with mechanical fastening devices to improve strength, durability, and vibration resistance. The easy property modification has resulted in a major line of fastening devices to enhance all areas of assembly. Advances have been made primarily based on new needs and performance requirements of industry. As new materials are developed, and as these materials are used in new designs and new concepts, anaerobic adhesives will be improved and available to improve product durability and reduce assembly cost.

REFERENCES

(1) Batson, R. E. and Tokarski, J, Technology of Threaded Fasteners, 1974 Loctite Corporation.

(2) Baccei, L. M.; and Murray, B. D.; "Anaerobic Adhesives, New Properties" SME Technical Paper Ad. 75-792 SME ASSEMBLEX October 1975.

(3) F. J. Campbell et. al. "Radiation Curable Resins" pp 79-92. Bascom, W. G. and Lockhart, L. B. (ed) High Performance Composites and Adhesives for V/STOL Aircraft, Second Annual Report. Naval Research Lab Memorandum Report 3721, Feb. 1978.

BLACK AND DECKER'S ANAEROBIC ADHESIVE SYSTEM FOR ASSEMBLING ARMATURES

William H. O'Connor
Black & Decker (U.S.) Inc.

INTRODUCTION

In the early part of this decade, Black & Decker pioneered the use of anaerobic adhesives to bond dissimilar materials into one structurally sound armature assembly. The resulting techniques have since been patented, and today are the backbone and competitive edge of one of the world's highest volume small motor manufacturing operations. The assembly includes an insulating tube bonded to the steel armature shaft, a stack of steel laminations bonded to each other and to the insulating tube, and a phenolic molded commutator bonded to the insulating tube. In the finished assembly, the insulating tube isolates the shaft from the other motor components. See Fig. 1.

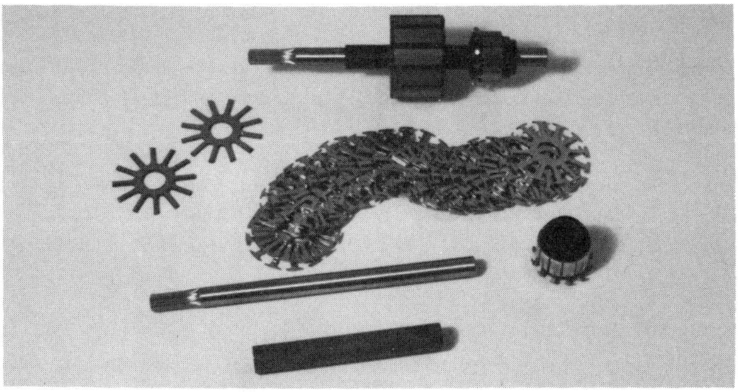

Fig. 1. Bonded Components in Armature Subassembly

BACKGROUND

In the late 1960's Black and Decker and the rest of the power tool industry decided to change to a double insulated construction on most of their home use products. In many respects, designing to this new concept improved the products and provided added convenience to the consumer. Instead of having the traditional three conductor plug and power supply cord, products were designed with a two conductor plug and cord. Within the tool we were now designing in two systems of electrical insulation instead of one. In order for this construction to be approved and listed by Underwriters' Laboratories, the second system of insulation, which is what this paper deals with, had to be completely functional if the primary insulation system failed.

REQUIREMENTS

This gets us into the specific requirements of the system being discussed. First are the safety requirements many of which were established by Underwriters' Laboratories. One of many tests conducted involved gradually increasing the load on an operating tool so the motor would burn up in a period of four to six hours. At the time of failure, it was not unusual to see motor temperatures in excess of 200 °C. In most cases, the primary insulation system was completely destroyed. However, it was required that the system, which in this case was the insulating tube, still be capable of holding a 1500 volt potential for one minute.

There were also requirements related to the physical integrity and reliability of the tool in everyday use. These requirements included:

- deceleration from 23,000 RPM to 0 in 300 milliseconds - 2,000 cycles
- 150 hours of continuous loaded use
- drop tests
- thermal cycling
- humidity tests
- abusive use tests

The other requirements were related to making good business decisions. We were obviously looking for the lowest possible cost and the most trouble free manufacturing process we could develop.

EARLY APPROACHES

Although we had been double insulating tools for years, it was on a relatively low volume basis compared to the millions of units a year we were then considering. The basic consideration was how to break up the continuous path of metal which runs from the rotating armature laminations to the armature shaft and on to the bearings and gears mounted in the aluminum gear case. Plastic gears and bearings mounted in plastic gear cases were very intriguing but not within our reach when considering requirements for heat dissipation, durability and life. Therefore, we had to fall back to an approach we were familiar with, and that was isolating the armature laminations from the armature shaft. The main problem here is that all mechanical metal to metal assembly techniques such as press fits, keys, knurls, splines, pins or brazing are eliminated. The method we had used previously was fixturing the laminations over the shaft and injection molding polyester insulating plastic into the clearance between the parts. This was a very slow and very labor intensive process resulting in a very strong assembly at a very high price.

Therefore our attention turned very quickly to insulating tubes. We looked at extruded phenolic tubes, glass tubes, fiberglass reinforced polyester tubes and paper tubes. We tried very hard to find a tube which could be slipped into the laminations and then expanded by pressing the armature shaft into place. Although this could be done, the tubes necessary to do it were very costly and the manufacturing process required very tight tolerances. Therefore the emphasis finally shifted to a paper tube insulator assembled between the shaft and laminations using anaerobic adhesives.

SYSTEM DEVELOPMENT

The development of this system required the joint efforts of product development and manufacturing engineers within Black and Decker and the tube supplier, adhesive supplier and assembly equipment designer and supplier outside of Black and Decker. I think it is important to note here that the coordinator of the project from the very beginning was the manufacturing engineer. This process was not engineered and then turned over to manufacturing. The person responsible for putting the process into production pulled the development through engineering. Along the same vein, the technician who tested thousands of prototype parts and helped to debug the first pieces of production assembly equipment moved to the plant and was the first supervisor over the bonding process when it started up.

In the early stages of developing this system there were a lot of problems, particularly with the sheer strength. Paper isn't the strongest material in the world and no one had ever used anaerobic adhesives to bond paper tubes to steel. So we were really on some very shaky ground. There were a lot of doubters, but the big incentive which kept us going on this project was the low cost and high potential for automation.

The basic test used in evaluating the system is a torque test performed by holding the stack of laminations at the 12 equally spaced slots and measuring the torque required to rotate the shaft. When we first started working with this system, it was encouraging to see torques of 3.92 newton-metres (35 inch-pounds). Today we operate to a minimum of 7.84 newton-metres (70 inch-pounds) with most readings being well over 11.2 newton-metres (100 inch-pounds).

During the development of the system, numerous materials, adhesives and processes were tried in the manufacturing of the tube itself. When all the problems were solved, it ended up being a resin impregnated kraft paper construction with a 1.02 millimetre (.040 inch) wall thickness. Each detail of the tube construction was painstakingly worked out using the tube suppliers expertise and Black and Decker's test facilities.

We went through similar experiences with the adhesive supplier. Each variable which could have an effect on the strength of the system was identified and evaluated. Various adhesives were tested in a search for the right formulation and viscosity. Torque was measured on hundreds of premeasured parts to determine the most ideal bond line dimensions. This information, along with process capability studies on each part, went into the establishment of tolerances for these parts. In a similar way the effects of the microfinish on armature shafts were studied and the results were used to establish a specification.

One of the most difficult problems was getting enough adhesive in the bond line between the shaft and paper tube and getting it properly cured. The solution to these problems required working closely with the adhesive vendor and the assembly machine designer. Methods were developed for automatically working the maximum amount of adhesive down the length of the tube where it interfaces with the shaft. On parts where we failed to get a good distribution of adhesive the entire length of the tube, the stack to shaft torque was all transmitted through a small section of the paper tube causing the tube to fail prematurely.

In evaluating the curing of the adhesive, the speed of heating as well as the temperature became critical. For instance if too much heat is applied too fast it will boil the adhesive back out of the bond line. On the other hand we were looking for the fastest reasonable cure time with enough heat supplied to fully cure the bond.

As we sought to increase the strength of this system the biggest problem was correctly identifying the weakest link and improving it. Was the failure due to the adhesive, not enough adhesive, the cure, the microfinish, dirty parts, too much clearance, not enough clearance, the tube or some other variable? If it was the tube, which of several variables in the tube caused the problem?

Even today the weak link drifts from one variable to another. The one thing that remains constant is a twice a shift torque study on each production line. If the torque starts to fall off, we can usually identify and correct the problem before it drops below the specified minimum.

THE PROCESS

The process for assembling the armatures was worked out by Black & Decker manufacturing engineers and the designer of the custom made machines. The adhesive supplier also interfaced with the machine designer since he was supplying the adhesive dispensing systems which are an integral part of the process. Within Black & Decker, a quality control engineer was included in the planning of the assembly process so we could incorporate as many checks as possible within the equipment. At the same time Development Engineering identified critical areas on the armature shaft such as the pinion and bearing journals so that special care could be taken not to damage them during assembly. Engineering also developed dimensioning systems for the components and subassemblies involved, taking into account where the positive stop locators were going to be on the assembly machines.

Throughout the development program, extreme emphasis was placed on cleanliness. All solvents and chemicals used in the plant were checked for their effect on the adhesive and all processes within our plant and our vendors' plants were checked for the presence of silicon or other substances which might inhibit the bonding process. In the interest of cleanliness the first step in the planning was to enclose the process in its own positive pressure room within the factory. This room, which is known as the bonding room, is free of oil mists and other contaminants which may be present in other areas of the plant. Before the shafts and laminations enter the room they are thoroughly cleaned in a vapor degreaser.

Before the tubes enter the bonding room, they are counter bored to a depth of 6.35 millimetres (.25 inches at each end). This is done to create more room for an adhesive resevoir between the tube and the shaft as we are trying to work the adhesive down into the bond line between the tube and the shaft.

At the beginning of the line, the armature laminations are loaded onto the equipment on 1 metre (1.1 yard) rods. The machine automatically selects the preset stack length and moves it into an alignment station. Here the 12 equally spaced slots of each lamination are aligned with the slots of the other laminations.

Since it would be very difficult to individually bond as many as 45 thin laminations to a paper tube we first unitize the laminations into one solid stack. See Fig. 2. This is done on an automatically fed 20 station rotary table. See Fig. 3. The parts are loaded, aligned, compressed and preheated by an induction heater which comes up around the part. Then a metal tube with a flanged end and flow of anaerobic adhesive down the OD is inserted in the ID of the stack of laminations. By capillary action, the adhesive is drawn out between each and every lamination. All of the adhesive flowing down the tube which does not go into the bonding process is recovered by a vacuum in the center of the tube and recycled. Before the parts are discharged from the machine, they pass through additional heat stations where the adhesive is cured. Unitizing the stack of laminations is strictly an aid to our manufacutring process. Once the armatures have been wound, this bond serves absolutely no purpose.

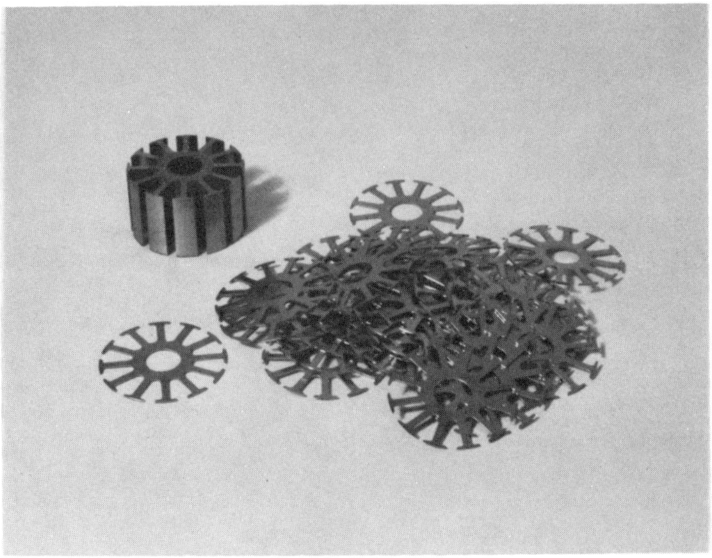

Fig. 2. Unitizing Laminations

Fig. 3. Unitizer

The first critical bond we come to is the tube to shaft assembly. See Fig. 4.

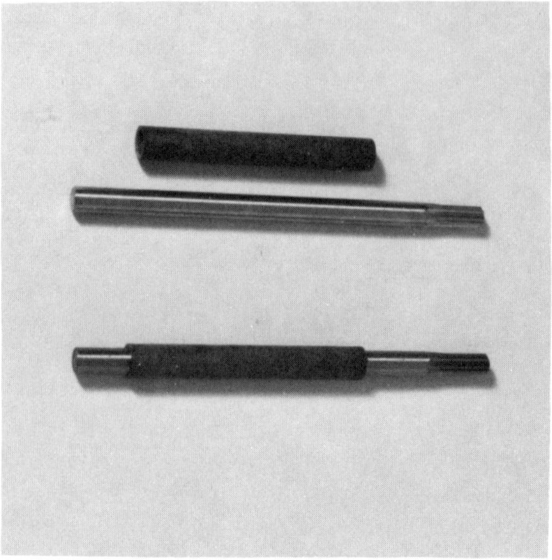

Fig. 4.
Tube to Shaft Assembly

This assembly is also done on a rotary table. However, the parts are loaded and off loaded by hand. The paper tubes are placed in vertical collets, and the armature shafts are dropped down into the paper tubes with the pinions up. As the parts progress around the machine, they come to a sensor which verifies that there are parts in each station. If, and only if, parts are present a needle comes in and applies anaerobic adhesive to the shaft as it is lifted and rotated by a spindle on the machine. Using a rotating up and down motion, the adhesive is worked down between the tube and shaft. In the next operation a bar comes down and accurately locates the shaft in relation to the tube. The parts then progress through several stations where induction heaters heat the steel shaft thereby kicking over the adhesive. Before the parts are removed from the machine, they pass through a station where a 132 newton (30 pound) push out test is performed. If for any reason a bond has not been formed between the two parts, a warning buzzer will sound and a red ink pad will move in and mark the defective part.

The final bonding machine assembles the unitized

stack of laminations to the shaft and tube assembly. See Fig. 5.

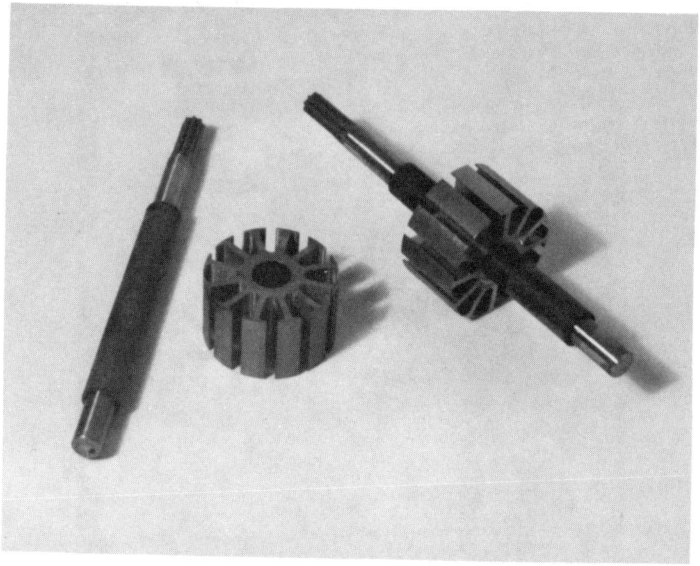

Fig. 5. Stack to Tube Assembly

This machine is very similar to the one just described and is also loaded and off loaded by hand. See Fig. 6. Here the unitized stack is dropped into a collet and the shaft/tube subassembly dropped into the stack's ID. Similar to the first machine the anaerobic adhesive is applied to the tube as the shaft/tube subassembly is lifted and rotated on the machine. By the same up and down rotating motion, the adhesive is worked down into the bond line between the paper tube and the unitized stack of laminations. After the final dimension is set, the adhesive in this assembly is also set by a series of induction heaters heating the steel parts. You may recall that on the first machine the shafts were loaded with the pinions up and the adhesive was worked into the shaft/tube bond line from the pinion end. On the bonder just described, the pinions are down and the opposite end of the shaft is exposed. Therefore we take this opportunity to add another shot of adhesive to the shaft/tube bond line at this end of the shaft just after the parts leave the induction heaters. There is enough residual heat in the shaft to draw in the adhesive and kick it over. Again on this machine a

Fig. 6. Bonding Machine

push out test is performed before the parts are off loaded.

After the last bonding machine, the parts used to roll down a track and pass through an ultrasonic cleaning operation to remove any residual adhesive from the bearing journals. Within the past year, it was determined that adhesive on the bearing journals, even in excess of what we normally see, would not clog the pores of our oil impregnated powder metal sleeve bearings or in any way affect their life. Therefore, the cleaning operation was eliminated.

The parts now progress directly to an end fibre placer which simultaneously presses an insulator onto each end of the impregnated paper tube and then locates them against the stack.

The part then rolls down a track to the commutator

placer. See. Fig. 7.

Fig. 7. Commutator to Tube Assembly

At the end of the track, the part is rotated while a thin bead of anaerobic adhesive is applied to the tube where the commutators, which have a phenolic ID, have been progressing from a feeder bowl through an oven. As they leave the oven they are picked up by a robot like arm and placed on the paper tube. The heat stored in the commutator is enough to kick over the adhesive. There is an automatic commutator bond check here which applies a force of 445 newtons (100 pounds) to the commutator and removes the part from the track if there is any commutator movement.

With the paper tube cut through to the shaft right next to the commutator, we can guarantee a minimum break away torque between the commutator and the paper tube of 5.6 newton-metres (50 inch-pounds) with the average value being about 11.2 newton-metres (100 inch-pounds). The failure mode here is almost always the tube to shaft bond.

The parts are now ready to be insulated and wound. After winding, there is one last adhesive application. A bead of anaerobic adhesive is applied to the end of the tube as the part is rotated. Then a mineral filled nylon fan with a stepped I.D. is pressed onto the steel shaft with a size to size to clearance fit over the tube where the adhesive is located. See Fig. 8.

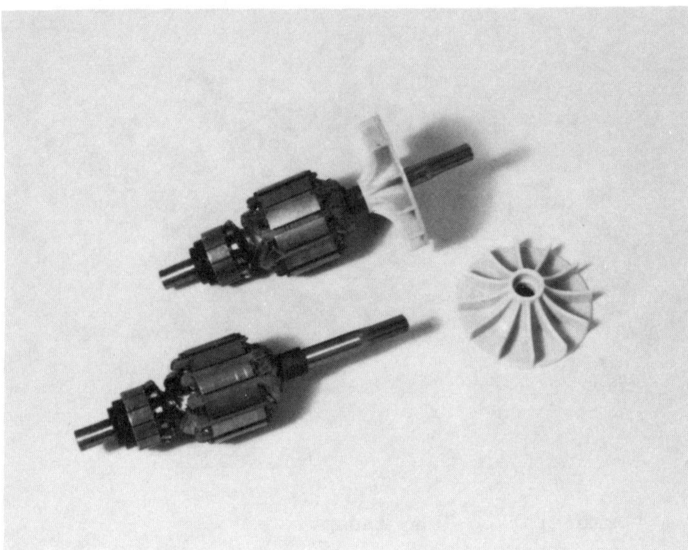

Fig. 8. Fan to Tube Assembly

As the armature is passing through a process to impregnate the coils with resin and cure the resin, the adhesive under the fan is also cured.

LIMITATIONS

Although this system has its limitations, it has proven to be completely reliable in all tools where the limitations have not been exceeded. All of the armatures we have ever used this system on have a diameter of 35.8 millimetres (1.41 inches) and produce up to 1/4 kilowatt (1/3 horsepower). The structural strength of the system and the requirements on the insulating tube itself would make it difficult to use this system on much larger diameters. The main limitation we have faced at the present diameter is related to the lamination stack length. Up to 28.58 millimetres (1 1/8 inches) we have complete reliability. These armatures are used in high vibration units such as jig saws and hammer drills without any problems. They are also used in hedge trimmers which are subjected to instantaneous stalls when they lock up on a chain link fence or a large branch. However when the armature gets up to 38.1 millimetres (1 1/2 inches) in length it becomes difficult to work enough adhesive into the bond lines and reliability starts to drop off. In such cases we have had to drop back and use the much stronger process where the laminations are molded to the shaft.

SUMMARY

The success of this Black & Decker design depends on the performance and reliability of anaerobic adhesives. Today millions of power tools in the field rely on the sheer strength of anaerobic adhesive bonds to transmit the torque developed in their motors to the armature shaft and the mechanical mechanisms being driven. The results have been excellent. Not only is the process reliable but it is still the lowest cost process available to accomplish our objectives. The end result is that this process contributes to our goal of providing the highest quality product at the lowest possible cost.

FORMED-IN-PLACE GASKETS

James P. King
International Harvester Company

The normal design requirements of a gasketed joint are that it be fluid tight under all foreseeable conditions, and sometimes, in addition, transmit mechanical loads. The conditions under which the joint must perform frequently include vibrations, thermal changes, transmitted load and pressure fluctuations, etc. In addition, the gasket must be chemically and physically resistant to the material it is sealing and the environment around it. In designing a gasketed joint, the engineer usually has a number of choices. Some of these are: bolt load, flange width and thickness, as well as type and dimension of the gasket. The resulting design must yield a net gasket load at all points of the flange sufficient to prevent blowout of the gasket under the applied pressure.

Another factor, in even a well designed gasketed joint, is the care taken in manufacturing and assembly. Over- or under-torquing of the bolts, improper flange dimensions, etc., can all affect joint reliability.

To meet these challenges, the designer can choose two basic types of gaskets. One is the conventional precut gasket made of some compressible material. The second is the formed-in-place liquid or paste type gasket.

The following are some of the advantages of the formed-in-place gaskets. First, one formed-in-place gasketing product can be used for many different applications, which results in reduction of factory inventory. Reduced inventory is also a convenience for in-field servicing of gasketed assemblies. Second, some of the filled types of gasket material seal effectively even with a slight gap between the flanges. Third, most of the formed-in-place gaskets have some adhesive strength. This feature increases the strength of the joint, and is particularly important in joints that are expected to carry mechanical loads as well as sealing. Finally, less costly machining operations can be used since formed-in-place gaskets can stand rougher surfaces.

Many methods of applying the formed-in-place gasket materials have been developed. Some of the more common of these are: (1) stencils which control the amount and placement of paste-like materials; (2) silk screen technique to place the less viscous types of gasket material. The coarseness of the "silk" screen medium is used to control the thickness of the gasket deposit; (3) application of a bead of material with a caulking gun either manually or air actuated; (4) depositing material from a squeeze bottle or tube. The type of applicator used depends on the quantity of production, size and complexity of the parts, and type of gasketing material.

The most important consideration for choosing formed-in-place gasketing is, the increased probability of producing reliable leakproof joints. Also, due to metal-to-metal contact, assembly strength and proper dimensional stack up, is more easily achieved.

SEALING BY IMPREGNATION

B. D. Baughn
Ford Motor Co.

INTRODUCTION

In the automotive industry there is great pressure to reduce the weight of all parts of the vehicle in order to meet government legislated Corporate Average Fuel Economy requirements. One way of achieving significant weight reductions is to substitute lightweight materials for some of the many heavy iron castings used in the drivetrain and chassis of the vehicle. A material commonly substituted for iron is aluminum. Aluminum offers a 63% weight reduction compared to cast iron. However, aluminum castings tend to be porous. In the past this porosity has prevented their use in applications requiring pressure tightness.

Die casting which is generally the most economical of the aluminum casting processes, produces parts that have interconnected porosity, such that they cannot normally be used in applications requiring pressure tightness. It is common in die cast parts which have undergone extensive machining to encounter leak test reject rates of 100%. There has developed a need for a reliable process which will seal porous die castings, not as a salvage or repair operation but as a regular production operation. Anaerobic impregnation has proven to be an effective alternative.

IMPREGNATION ALTERNATIVES

There are three basic impregnation processes which are available on the market today.

Sodium Silicate impregnation, the so called "water glass" method, has been used for many years as a salvage operation. The yields have never been very good and the material is limited in its pressure and temperature ranges. The Sodium Silicate system is not generally considered acceptable in an application when very high yields are required.

Anaerobic impregnation and thermo-setting polyester impregnation are both effective processes for the impregnation of aluminum die castings. The advantages and disadvantages of each will be described later.

PROCESS DESCRIPTION

The anaerobic impregnation process consists of the following steps. First clean, dry, and temperature stabilized parts must be placed in the auto-clave. A vacuum must be drawn to remove all air from the pores in the parts. Then, an amount of the impregnation resin sufficient to cover the parts is introduced into the tank and the tank is pressurized to force the resin into the pores.

After the vacuum/pressure cycle, the parts are removed from the tank and undergo a spinning operation to remove excess resin from the surface of the parts. The high cost of the resin makes it important that as much resin as possible be returned to the tank for reuse.

The next step is a cold water/detergent wash to remove all resin from surface of the part without removing it from the pores. This is followed with a cold water catalyst rinse and a final warm water rinse. The catalyst and heat are used in the last two operations to accelerate cure of the resin. Although the resin will cure without the catalyst or elevated temperature, the time delay between impregnation and subsequent pressure testing can be significantly reduced.

IMPREGNATION ALTERNATIVES (Cont'd)

The thermosetting polyester impregnation system is very similar to the anaerobic process. The polyester system generally does not use a spin operation following the autoclave. The lower resin cost and higher resin viscosity make the spin operation not cost effective.

Following removal from the autoclave, the parts are washed to remove resin from the surface and then cured. The cure is generally performed in a hot oil bath.

CONTROL OF THE ANAEROBIC IMPREGNATION PROCESS

Control of the anaerobic impregnation process is not simple. However, experience has shown that the process can be controlled in a production environment if proper controls are used.

The particular areas that require careful attention are as follows:

- High resin temperature contributes to instability of the resin. A refrigeration system is generally used to keep the resin in the preferred 40-50° F range. The reliability of the cooling system is usually insured with a back-up refrigeration system.
- A continuous flow of air through the resin must be maintained. Anaerobic resins will cure without a sufficient amount of air available.

The failure to control the two variables noted above can result in the entire tank of resin curing in the autoclave. The loss of as much as 600 gallons of resin and the damage to the system can by very costly. The air flow and refrigeration must be maintained continuously after the system is activated. It cannot be shut down during off-shift periods.

- Careful attention must also be used to prevent contamination of the resin. The system is tolerant to a reasonable level of dirt, water and other contaminants. The contaminants may affect the ability to seal, the stability of the uncured resin and the mechanical properties of the cured resin.

CHOOSING THE RIGHT SYTEM

Choosing the right system for a particular application can be quite difficult. Quantifying the costs and benefits of the system can only be done with a carefully controlled trial run.

The two factors that are difficult to quantify are resin useage and yield. Resin useage is affected by viscosity, part orientation during the process and the effectiveness of the spin cycle. Resin useage is a major component of the cost of impregnation.

Yield is also difficult to quantify because it is highly dependent on the nature of the porosity and on the cleanliness of the part.

DISPOSAL OF WASTES

A special word of caution, some impregnation processes can present waste disposal problems. These problems will vary with the local environmental protection laws and the particular type of impregnation systems being used.

CONCLUSIONS

Impregnation can be an effective tool to seal porous parts. The process is useful not only to salvage parts which have been rejected but also as a regular production process to seal porous parts. Impregnation can be the key to finding the optimum design.

CORROSION PROTECTION — I:

MATERIALS AND DESIGN IN HIGHWAY AND OFF-HIGHWAY VEHICLES

PROGRESS IN THE CORROSION PROTECTION OF AUTOMOTIVE COMPONENTS

Ronald S. Michalak
Engineering Group Manager
Fisher Body Division
General Motors

Introduction

Corrosion is one of the major problems facing not only the automotive industry, but the makers and users of almost every product in our modern society. One recent estimate indicates that corrosion of all types cost the American public over thirty billion dollars annually.

Corrosion protection is a battle that we at G.M. are intensely fighting and winning. Winning in that we are confident enough of our product to offer an extended warranty against performance on all 1980 Model G.M. cars and trucks sold in the United States & Canada.

This paper covers the broad spectrum of automotive protection. It will do this by discussing the basic types of automotive corrosion, follow with four categories of potential solutions, and conclude with our current methods of evaluation and detailed description of one of G.M.'s latest packages.

Definition

Corrosion and the job of combatting it in its many forms is a complex task, one that requires definition. At General Motors, we break corrosion down into two basic types: Perforation and Cosmetic.

Environment

Perforation Corrosion most generally starts from inside out, although not always, and is most prevalent in the northeastern sections of the United States and Canada. While this represents about 25% of the U.S. land mass, the heavy population concentration in these regions means that nearly 50% of the vehicle population is subject to this form of corrosion. The primary cause for corrosion in this zone is the use of road deicing salts which have more than doubled in the last ten years. Some municipalities compound the severity of the road environment by adding calcium chloride or slag to the salt.

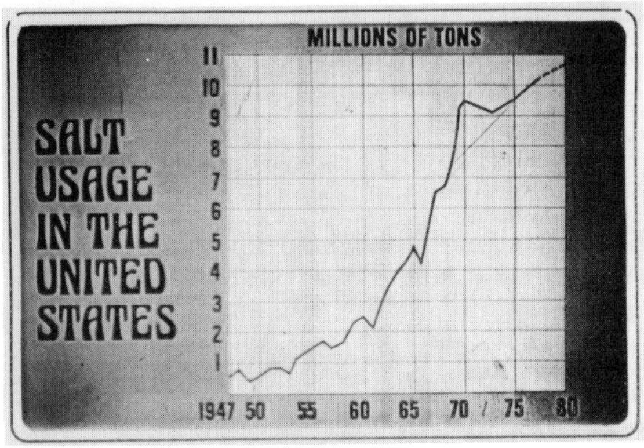

Fig 1

Road Salt Usage

This chart (Fig. I) of salt usage taken from information published by the salt institute begins to explain why corrosion is prevalent in northern regions of the country.

Now, none of us would recommend an elimination of deicing salts for road usage. . . for without deicing salts, our winter time mobility would be lost.

Therefore, we believe you can expect to see the use of salt continue to grow, which means that we're dealing with a moving target.

Perforation corrosion is most insidious in automotive products. It most often starts in places where the customer does not know he has a problem until it's too late, such as this. (Fig. 2)

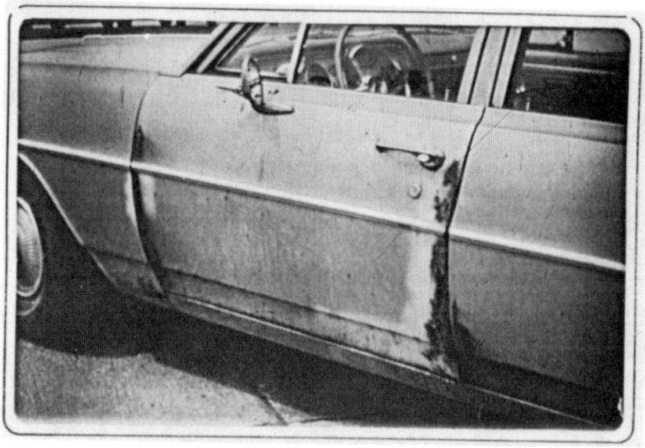

Fig 2 Example of Perforation Corrosion

Our objective is to try to eliminate corrosion of this kind.

The second category of corrosion, which we call cosmetic, gets its name since it usually can be seen in its early stages by the customer. It is prevalent, but not limited to coastal areas where humidity is high and salt spray from the ocean breezes cover our products. (Fig.3)

Fig 3 Example of Cosmetic Corrosion

It can also be seen, however, in other areas and, in particular, those that use deicing salts.

It gets its earliest start anywhere the protective coating has been misapplied or damaged either in installation or service. Thin paint in hard to spray areas, sealer skips, plant and parking lot damage, and stone bruises are among the chief sources.

An aid to minimizing plant damage is employe awareness. We distribute a booklet to our assembly plant personnel to encourage them to practice damage avoidance techniques.

Solution

Now, let's turn to our four cateogires of potential solutions. . . Design, Materials, Coatings and Sealing Systems

I will look at each one of these and highlight some of the progress that has been made and the top challenges still in need of solution. Progress has been signivicant. Testing and analysis tells us that our current models match up well as fighters of the two corrosion enemies.

Design

First design can be used to minimize corrosion three ways. We can eliminate vulnerable surfaces and crevices where possible.

An example of this is the vehicle underbody. We relocated the major cross-bars of the floor pan assembly to the topside of the pan from the exposed underside. (Fig. 4)

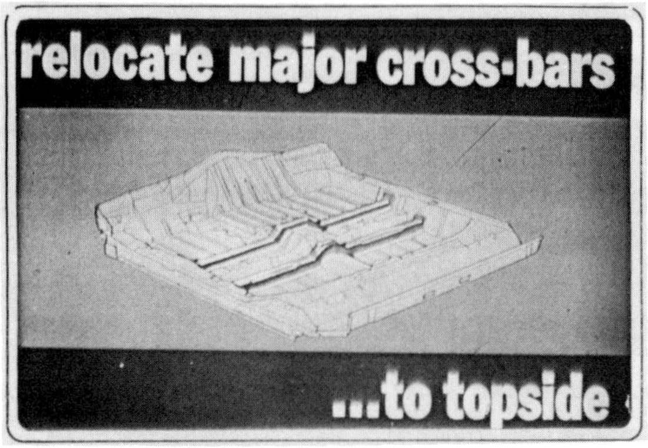

Fig 4

Example of Vehicle Underbody

This eliminated the boxed-in sections where poultice and crevice corrosion can take an early toll.

We first took this approach with our 1973 intermediate sized cars and this has been a design criterion in every new body program since.

The redesigned luxury series for 1979 & 1980 feature another corrosion fighting design innovation which is a first That is, the flush glass windshield installation. (Fig. 5)

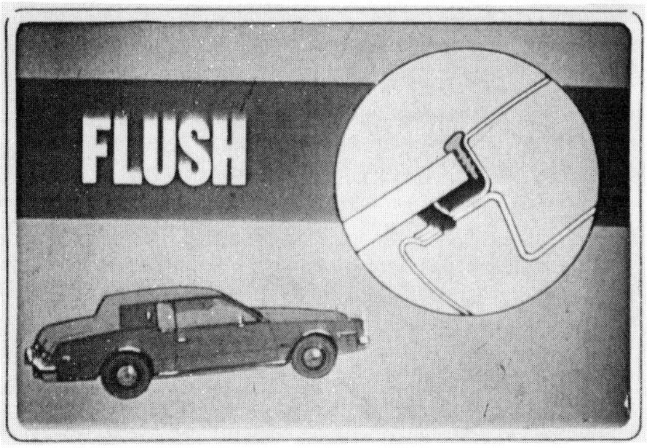

Fig 5 New Design Flush Glass

The traditional metal windshield molding with its weld studs and attaching clips has been replaced by a thin plastic molding which is bonded into place. Not only do we eliminate the corrosion prone crevice between the metal molding and body, but also the potential water pocket beneath the molding and chipped paint during installation.

This also provides a sleek, aerodynamic appearance as modern and fresh as tomorrow.

Another way that design can minimize corrosion is by keeping the environment away from interior metal structures. Elimination of holes in the exterior panels of an automobile is a top objective in this category.

The development of adhesively-bonded moldings, scripts and emblems that don't require holes for attachment purposes is one of the best examples of this corrosion-fighting design approach. (Fig. 6)

Fig 6 Adhesive Bonding

This design change also eliminates the potential for cosmetic corrosion that can occur at metal molding clip sites.

The third design approach is to provide good access for protective coating application and subsequent drainage or aeration of the area in service when exposure to a harsh environment cannot be avoided.

Our redesigned front fenders are an example of this approach. The spacing between the fender and the lower rear reinforcement has been increased and possible poultice collecting weld flanges repositioned to promote positive drainage. (Fig. 7)

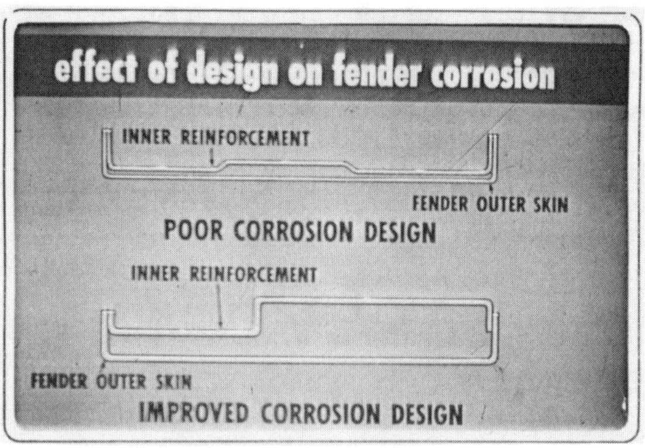

Fig 7 Front to Support Spacing

While we have made substantial progress in designing-out corrosion, many areas of concern need more work. Let me identify just two design challenges where we can use some improvement.

Take the case of the screw head recess which can be mutilated during installation. This destroys the protective finish and results in early cosmetic corrosion.

While improvements are being realized thru imporved finishes, use of stainless steel, and imporved recess designs for selected locations, we need a universal solution for the regular drive screws used in most applications.

Some ask why we just don't use stainless steel screws for all these applications. One reason is that they are not hard enough to give us relizable drive characteristics day-in and day-out on the line. What we need is a real innovative breakthrough in this area.

Another area is protection of operating mechanisms. What is the best way to redesign or re-engineer operating mechanisms to eliminate cosmetic corrosion that can occur at moving joints and wear surfaces, such as, in the case of door hinges?

Materials

Going on, our second basic corrosion fighter is materials. The most successful materials to date for this purpose have been the zinc-coated steels. In fact we have seen a dramatic rise in the use of precoated steel in automotive construction (from about 5% to about one-third of the steel tonnage used at Fisher Body).

Two-side hot-dip galvanize steel has been used successfully to combat corrosion in the earliest automotive applications at G.M. . . . rocker inner and outer panels, and quarter panel fillers since 1963.

Since the mid 1970's its use has been expanding to include wheelhouse outer panels, rear compartment pans and other panels where the environment is most severe.

However, two-side hot-dipped galvanized steel has been generally unsuited for major exterior panels such as doors, quarters, fenders from a number of standpoints, including: forming, assembling, finishing, and repairing.

Several modified two-side coated products have been used on exterior panel applications with limited success, but either unfavorable long-term field exposure or manufacturers' supply problems have caused a change in their use.

What was really needed was a one-side coated product for these major exterior panels, I congratulate the coatings and steel industries for innovating a family of one-side products in response to this need.

Zincrometal was the first, high volume one-side zinc-coated steel utilized for major exterior automotive panels. Early zincrometal required a lot of development work, but it was well worth the effort. It is being formed into the full range of exterior automotive panels, doors, quarters, fenders, and deck lids.

Zincrometal was joined by one-side electrogalvanize sheet steel in 1977 when facilities capable of providing high tonnage shipments came on stream. Initial use of this new capacity was in G.M.'s 1978 intermediate-size car quarter panels with other quarter panels being added to the list as availavility increased.

To challenge zincrometal and electrogalvanize, two steel companies have developed one-side, hot dipped galvanize automotive sheet products produced by two different, novel methods.

These products provide added volume of one-side coated steel to supplement and compete with zincrometal and electrogalvanize steel in the automotive market place.

This leads me to our latest quest in zinc coated products. If you noted earlier I stated that two side galvanized steels have been generally unsuited for major exterior panels. Our most recent request of the steel industry has been to come up with a two-side zinc coated steel with a surface acceptable for major exterior panel use.

A number of differentially coated material of various alloy content have been suggested for use on one-side coated steels. At this point it is hard to predict just how they will stack up against the one-side coated steel.

They all offer a high level fo corrosion protection for the steel substrate, and as we learn more about how these materials form, join, look and perform, optimum usage patterns will emerge.

Precoated steel is not without challengers, especially

today when mass reduction is a primary objective. A note about one of these challengers, aluminum, is in order. Aluminum has found use in various applications for exterior panels, with hoods, tulip panels and deck lids being primary applications. Aluminum corrosion resistance has been rated as excellent, but coupling with steel, in corrosion environments, can lead to serious galvanically-induced corrosion problems which must be carefully avoided. Could this lead to some interesting aluminum-galvanized construction possibilities in the future? Possibly, however, due to the massive price increases we have been experiencing recently, up to 40% in the last two years for certain alloys.

The use of aluminum sheet has been slowed considerably, so much for precoated metal and aluminum.

Coatings

Now let's take a quick look at fighting corrosion with coatings applied by the automotive manufacturer, which, as technology advancements are made, are getting their second wind and becoming serious challengers to zinc-coated steel for improved corrosion protection.

This includes both the basic paint system and augmentation materials such as zinc-rich primers, deadeners, waxes, and MOPB coatings.

These third corrosion protection alternatives appear to be especially effective if we think about them coupled with the fruits of the "clean steel" program.

The painters have been telling us for years that different pieces of steel clean and phosphate differently which can result in variable corrosion performance of the final painted product, and they are right. This slide (fig. 8) shows the paint creepback corrosion results of graveled and scribed salt spray test panels representing the edge-center-edge slits of a singel automotive sheet steel coil. We call this challenge our "clean steel" program. Our goal is to have production steel as produced by the mills and processed by fabricating plants, reach the paint shop with an optimum steel surface condition conducive to the formation of a quality phospate coating.

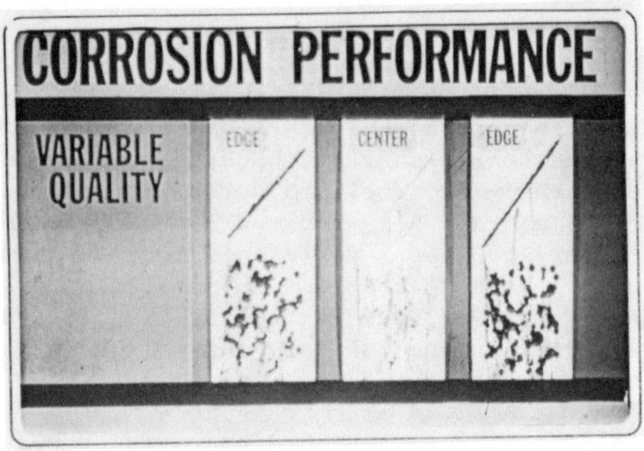

Fig 8 Corrosion Results on Steel Not
Processed for Clean Steel

The mills and our efforts during the pastyear have resulted in marked improvement in the phosphating quality of production steel with more and more steel coils showing uniformly good performance as shown here (fig. 9). We are looking toward further improvement to achieve an ultimate steel surface condition that optimizes as-painted corrosion resistance and when we find it, perhaps it will be good enough to protect some parts of the vehicle without the need of precoated steel.

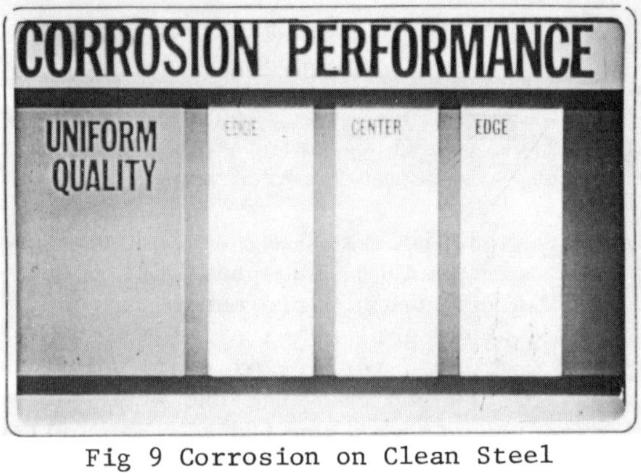

Fig 9 Corrosion on Clean Steel

The capability of the basic paint system (phosphate and primer, with surfacer and topcoat on appearanc surfaces) has been significantly improved with the development of the electrodeposition primer system, at G.M., we call it ELPO.

We installed our first electrodeposition primer tank at a body assembly plant in 1966 and have been expanding this technology ever since.

At present, we have 43 ELPO tanks priming passenger car and truck bodies, hoods, fenders, and small parts. The original ELPO primers were applied with the body as the Anode and were called ANODIC ELPO.

Even though Anodic Elpo showed many desirable attributes during the early and mid-70's, we felt that further refinements might still be made in corrosion resistance.

The paint industry stepped up to the task with the development of a cathodic Elpo primer.

The cathodic elpo shows a marked improvement in product durability (fig.10) so much so that we have converted most of our elpo tanks to this technology in the past two years with the remainder scheduled for conversion in the near future.

Fig 10 Cathodic ELPO Improvement

Eventually, as we see it now, all primer processes in G.M. plants will be cathodic elpo by 1982.

The development of cathodic elpo to replace existing anodic paints is expected to boost the coating system's performance especially in the most corrosion-prone, interior areas based on testing to date. However, it is premature to make any predictions as to what the percentage improvement will actually be without extensive field experience.

The body paint system consisting of zinc phosphate, primer (Elpo or spray), sealer or surfacer, and color topcoat provide an effective corrosion protection system on most exterior body surfaces. The same cannot be said for paint systems on interior surface, consisting of phosphate and prime only, for they require extensive augmentation in the form of pre-coated steels or augmentation materials in penetration-susceptible areas of the vehicle.

These include areas of poor or limited access-like quarter panel pockets, door cavities, boxed underbody sections, or abrasion areas such as wheelhouses where stone throw is severe (fig. 11). In some body locations where the environment can be expected to be most severe, two or three layers of protection are common, including pre-coated steel or other pre-priming operations.

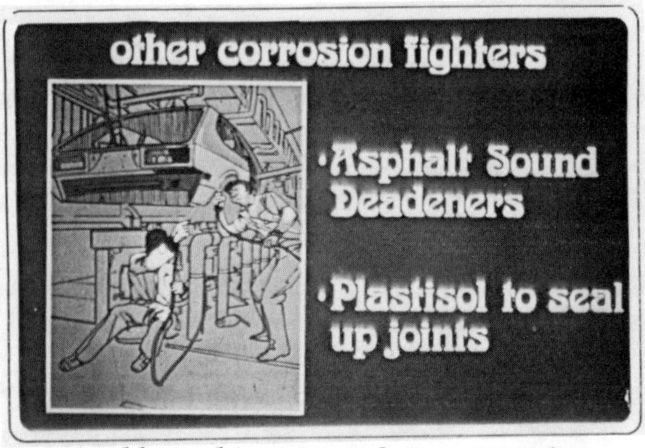

Fig 11 Application of Augmentation Coatings

Areas where moisture can be expected to accumulate for long periods are coated with aluminimized wax or MOPB coatings as a final operation.

Another example of augmentated coatings that are receiving considerable attention recently is a layer of chip-resistant plastisol applied over the primer prior to application of color topcoat on lower body surfaces subject to stone bruises. (fig. 12)

Fig 12 Application of Plastisol

Its initial use was on the 1976 Chevette rear quarter panel behind the rear wheels. We extended this protection to the same area of many 1980 models.

We expect to increase its use as we learn how to incorporate it into more of our paint shops, and confirm its product performance in the field.

Though these have been monumental achievements in our basic paint and augmentation programs I can still think of a couple major challenges we face in the coating areas, these are:

1. Improved phosphate coating systems, and

2. Improvements in repair procedures that make it easier to practically attain the level of corrosion resistance imparted by the original production coating process.

We need some real innovation in these areas to maintain the gains we have achieved as a result of our conversion to cathodic Elpo.

Sealers

The final corrosion protection alternative is sealers.

The environmental demand placed on interior body protection systems can be minimized or eliminated by effective sealing. In addition they provide protection against crevice and/or galvanic corrosion between dissimilar metals at the host of body joints and seams.

The current challenge in sealing is primarily related to appearance of plastisol-sealed, visible hems and joints. Flow-on application is inherently operator-sensitive. What's needed is a totally new approach, like a controlled spray-on application that can seal wide variations in joint conditions, produce a smooth, flawless joint appearance, and be emissions free, of course. That's a tall order, but well worth the challenge.

Performance

Stepping back for a moment and looking at the whole vehicle, it is apparent that dealing with corrosion protection must be a total systems basis.

Since the susceptibility of different areas of the car varies with the type of environmental condition experienced, our task is to design and protect the metal against the most severe conditions normally expected by each distinct area of the car. The problem we face is how to accurately determine the corrosion modes that our cars can experience, which designs and coatings are best suited for the job, and just what is the environment that needs to be protected against. It would be an easy straight-forward task to combat automotive corrosion if we knew that a given test or series of tests could predict service performance without a doubt, so that each material or system could be directly compared without any alibis, test prejudice, etc. Well, in the automotive business, we haven't seen one single test that can do that. But we keep looking.

We have several alternatives, however, which we use to help us to obtain the best answers available. Accelerated laboratory tests can give us a feel for a material or system in the shortest possible time by comparison to known systems, however, there is a constant need to confirm the findings with environmental exposures at static sites like the seashore, over-the-road fleet tests in various areas of the country, or under full vehicle, accelerated conditions at the Proving Grounds.

In recent years, the trend has been to develop and place more confidence in cyclic tests in which the system is alternatively exposed to a series of individual test conditions, like heat, cold, salt spray, etc. to more precisely duplicate severe conditions.

The G.M. Proving Grounds has been using such a cyclic full

vehicle test for some time. The test includes a four hour contamination drive schedule through salt splash, salt spray, gravel, mud, dust, highspeed driving, and exposure to humidity soak. We are leaning more and more heavily on this test to give us overall design direction for corrosion protection.

Of course, this testing is for naught unless it correlates with field results. Information from all tests plus service results serve together as the basic source of knowledge on body corrosion, means for its control, and on which final engineering judgements are made regarding material design and costing selections.

Product Package

To see just how the corrosion package goes together today' let's take a look at the new 1980 compact-size G.M. cars, (Fig. 13) which represent the results of our latest effort to meet the corrosion challenge.

Fig 13 1980 G.M. X-Body

Precoated Metals (Fig. 14)

While rocker inner and outer panels and quarter filler panels have been two-side hot-dip galvanize for many years, the entire rear compartment pan and reinforcement assembly is new for 1980. In addition, the wheelhouse panels and certain front end members and rear end panels have also been specified in G-90 galvanize.

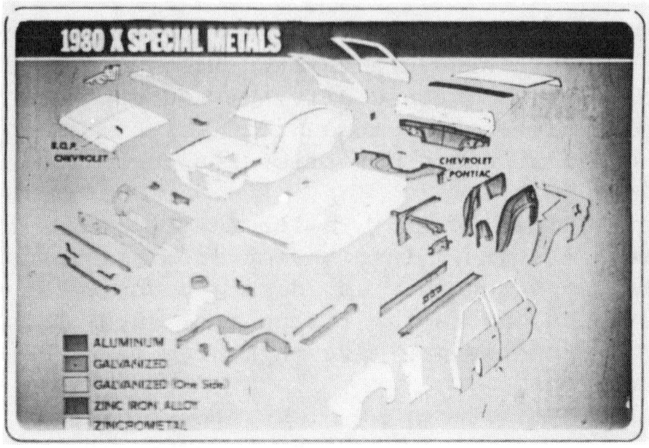

Fig 14 1980 X-Body Pre-Coated Metal

The quarter panels on these new compacts are the major use for one-side electro-galvanized steel. This product was also specified for the door outer panel application. However, zincrometal had to be used due to lack of one-side electro or hot-dip galvanize capacity. The front end plenum chamber is likewise one-side galvanize.

Zincrometal is specified for the less corrosion susceptible areas such as deck lid and hatch lid outer panels and upper shroud panel.

Pre-Prime Coatings

Zinc-rich preprime is used liberally in addition to precoated steels at the lower perimeter of the door inner panel and other selective locations to complete the "pre-coat" package for this new vehicle line.

The entire body is primed an anti-chip plastisol coating is applied prior to primer-surfacer and topcoat.

Although all body joints are sealed visible joints are cosmetically critical and special attention is given to these.

A new secondary hot melt seal in addition to body shop weld through sealer is specified for the wheelhouse flange to reduce potential water leaks in this area. Drainage has been improved in the bootleg pocket as well. Deadener is continued in high abrasion areas in the wheelhouse and the compartment pan directly behind the wheels.

After full paint, selected corrosion susceptible areas receive further augmentation in the form of wax coating to complete the basic body protection system.

Summary

In summary, I think you can see that strides are being made toward reducing the effects of corrosion on the appearance and usability of today's and tommorow's automobiles. The design consideration, basic material selection and applied coating all contribute to a successful package. There are, however, numerous challenges facing each of us in the battle against corrosion, and only though continuing advances will we win the battle against corrosion.

ZINC-RICH PRECOATED STEEL

Griff W. Froman
Inland Steel Company

With the cost from the corrosion of cars and trucks in the United States alone estimated at $5.5 billion annually, corrosion prevention is probably the main task of materials engineers today. The most widely used methods for corrosion protection of steel involve coating the substrate with some sort of protective film, to either isolate the steel surface from the corrosive environment, or to sacrificially protect the steel by the preferential corrosion of the coating system.

Low-alloy plain carbon steels are commonly coated with aluminum, chromium, nickel or similar highly corrosion resistant metals to prevent the oxidation of the steel substrate. Such barrier films are highly effective corrosion preventive coatings unless the continuity of the film is disrupted during forming or by abrasion. In this case, the more noble coating acts as a large cathode of an electrolytic cell and in an aggressive environment actually accelerates the corrosion of the anodic steel base.

Galvanized coatings afford protection of the steel substrate by sacrificial means. Such films are anodic with respect to the steel and corrode in preference to the substrate. The degree of corrosion protection is directly related to the zinc film thickness and is only effective, of course, while metallic zinc is available.

By far the most commonly used protective coatings are based on the barrier film properties of paint systems. The total value of paints and lacquers produced in the United States amounts to about $2.5 billion annually, half of which is estimated to be used for corrosion prevention. All present-day paints are permeable in some degree to water and oxygen. Some resin systems are less permeable than others and, in general, their better performance as a diffusion barrier applies only to adherent multiple-coat applications that effectively seal up pores and other defects.

Zinc-rich paints represent an attempt to combine the sacrificial corrosion protection of zinc with the barrier film protection of paints. Finely divided metallic zinc particles are embedded in a resin matrix and even though the particles are covered by a thin film of resin a relatively good conductive path is produced by the point-to-point contact of the zinc dust. The electrical conductivity of this system allows the zinc to behave anodically to the steel substrate when a corrosive environment eventually permeates the paint film. Zinc pigmentation can be incorporated into either organic or inorganic binder systems. The inorganic films are usually quite brittle and are therefore applied as spray-on coatings to pre-assembled structures, whereas the organic zinc-rich paints exhibit the adhesion and flexibility required for prepainted applications.

Currently, U.S. coil coating facilities are capable of supplying approximately three million tons of precoated steel coils annually and several new coating lines are now under construction. Modern coil coating lines provide uniformity in coating film control and versatility in processing capabilities unique to the coatings industry. The schematic of a coil coating line, shown below, demonstrates the typical coil processing sequences available. Typically, steel coils are subjected to a variety of cleaning, brushing, and pretreatment stages prior to the application of the paint films. The coating rolls are normally operated turning opposite to strip travel with coating thickness controlled by roll speed, and film uniformity maintained by contact pressure. Curing oven capacity has been increasing so as to now accommodate line speeds over 500 feet per minute for steel strip close to 70 inches wide.

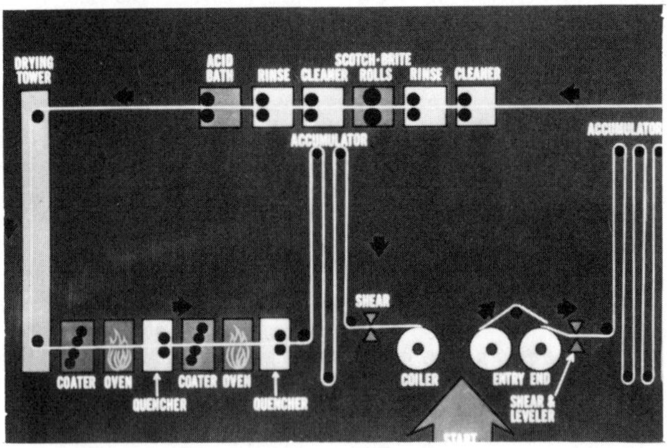

Fig. 1. Schematic of Coil Coating Line

This type of coil line has the capabilities for many prepainting operations which assures strip cleanliness and provides a variety of possible metal surface preparations. The coating rolls can be operated to paint both sides or only one side of the steel strip and can even be designed to paint only selected areas with no adverse effects on the uncoated steel surface characteristics.

A precoated product for use in a large market such as automotive must meet several important property requirements. Adequate corrosion resistance is, of course, very important but the coating must also have the adhesion and flexibility to withstand deformation when formed into subassembled parts by brake-press forming, roll forming, or stamping operations. Since the end user must further assemble these coated parts, the coating system must be weldable. The coating must also be compatible with conventional topcoat or exterior finish coatings. Zinc-rich paints have the capacity to meet these important material requirements.

A typical zinc-rich paint is produced by mixing very fine metallic zinc powder into a resin system to develop a complex film, as shown in the accompanying photomicrographs. The resin appears to cover and adhere to the zinc particles and thus binds them to the substrate as an integral film. A zinc dust size distribution is evident with a maximum particle size typically around 8 to 12 μm. The cross section micrograph shows the particle distribution to be also quite random throughout the thickness of the coating. The smaller particles tend to improve the packing efficiency of the dust thus increasing the number of contact points between particles and the overall contact area with the substrate. The packing efficiency is a function of particle size and concentration, which in turn affects the paint film conductivity.

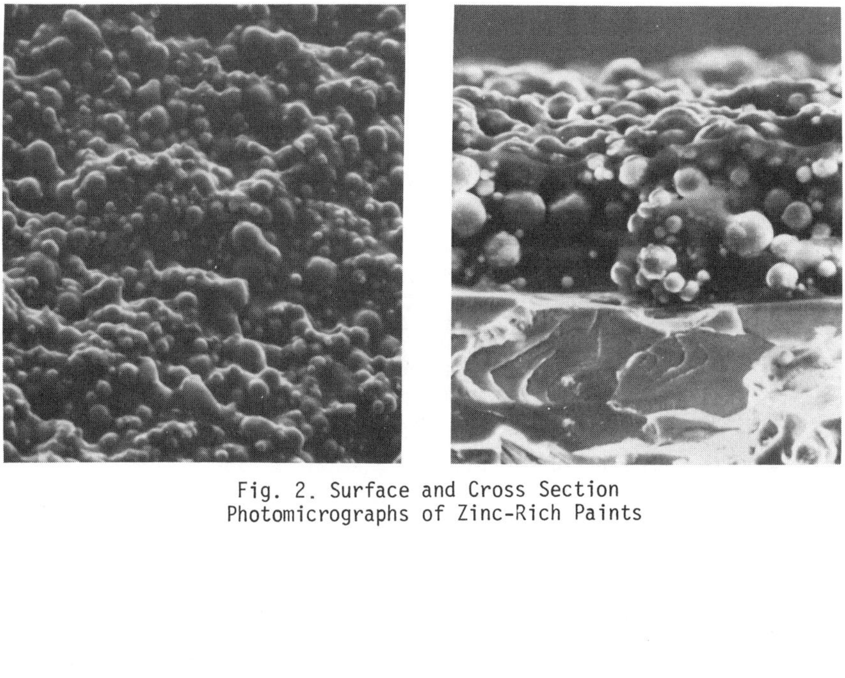

Fig. 2. Surface and Cross Section Photomicrographs of Zinc-Rich Paints

The most important factor contributing to the desirable performance properties of zinc-rich paint systems is the ability of the film to conduct electrical current. The resistivity of zinc-rich films as a function of zinc dust loading is shown below. Most zinc-rich systems operate at zinc loadings of between 80% and 95% by weight. This degree of metallic zinc pigment loading produces adequate particle contact which provides the electrically conductive paths for desired film conductivity.

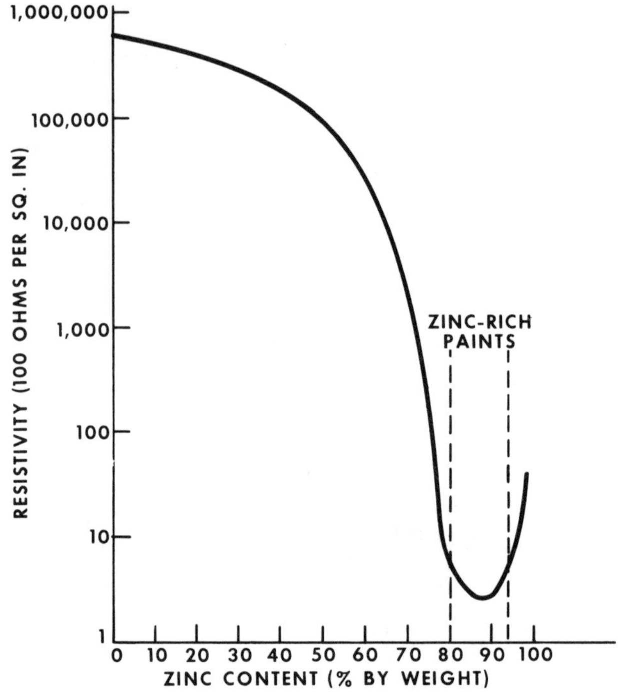

Fig. 3. Resistivity of Zinc-Rich Paints as a Function of Zinc Content

With such electrically conductive paths present throughout the zinc-rich matrix this painted product has been shown to be weldable, a property which expands the application possibilities over conventional painted coil products. The attached weldability lobes of a typical zinc-rich coated steel, a plot of weld cycles versus applied current, show a rather large range of resistance spot welding conditions available to achieve an acceptable weld button.

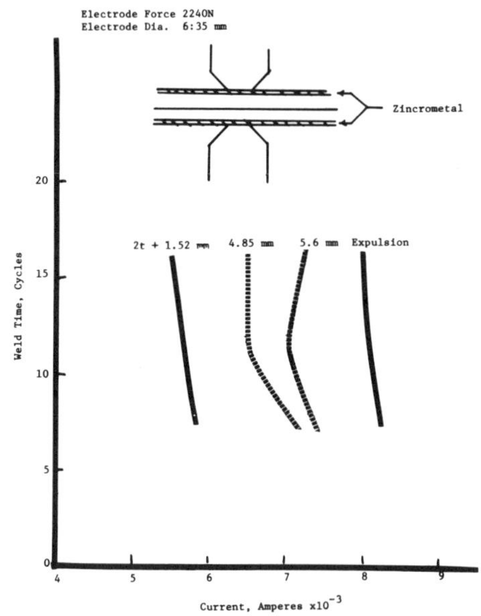

Fig. 4. Weldability Lobe
for Zinc-Rich Coated Steel

The weldability of a zinc-rich precoated system is also a function of the pretreatments or metal surface preparations required for any particular application. If a relatively non-conductive pretreatment such as a phosphate conversion coating is applied prior to the application of a zinc-rich paint, the resistivity of the entire precoated product is increased. Some pretreatment systems have been developed in which zinc, carbon, or other such electrically conductive pigments have been incorporated, thus maintaining an electrical coupling between the paint film and the substrate. The weldability lobes for the different surface preparations reflect the system's ability to conduct electricity, i.e., broader welding parameters are possible for the new conductive films, and a rather limited welding range is typical of the more insulative coating systems.

The electrical conductivity of the zinc-rich system also has an effect on the corrosion protective mechanism provided by the coating. The highly conductive systems exhibit a greater degree of white zinc corrosion products on the surface of test panels exposed to 5% salt spray exposure than do systems with lower conductivity. The prevention of red rust corrosion from the iron substrate at scribed or severely formed areas is also more evident with the more conductive systems. The more integral electrical coupling of the metallic zinc dust in the paint to the steel substrate allows for a greater degree of anodic or sacrificial protection. The less conductive systems provide a lesser degree of sacrificial protection at exposed areas but yet still provide general barrier film protection.

Anodic polarization measurements, shown for some typical materials in Figure 5, illustrate the different corrosion mechanisms. Conductive zinc-rich systems composed of paints applied directly to the steel substrate, or in conjunction with a conductive pretreatment, exhibit similar passivation behavior when immersed in 0.5 N sodium sulfate solution. The same zinc-rich paint applied over a phosphate conversion coating exhibits a behavior expected of an insulating layer with limited electrolyte permeability. The open circuit potentials of the conductive systems are very similar to those of hot-dip galvanized coatings, and the amount of current flow at 100 mv over open circuit may indicate the extent of local corrosion cells set up within the system by the contact between the dissimilar metals. The lower current values at open circuit for the more electrically resistant system demonstrates the lack of these sacrificial corrosion reactions. It is hoped that further work with such polarization techniques would lend more insight to the corrosion mechanisms of zinc-rich systems, and aid in determining long-term corrosion protection.

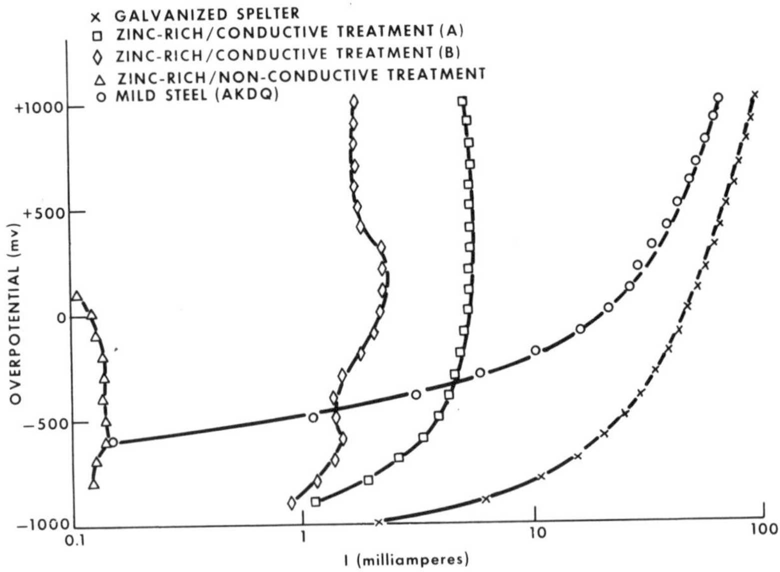

Fig. 5. Anodic Polarization Curves of Cold-Rolled, Galvanized and Zinc-Rich Coated Steels

The comparison of zinc-rich paints to hot-dip galvanized products is unavoidable since both systems incorporate metallic zinc to protect an iron substrate. Such comparisons however can be very misleading because the products are substantially different in many respects. Zinc-rich paints usually contain only about one quarter as much zinc per unit area as does 1-1/4 oz. (G.90) hot-dip galvanized steel. The paint offers barrier film properties, incorporates inhibitive pigments which passivate the zinc dust and steel substrate, and zinc-rich paints are not designed to withstand exposure to ultraviolet light sources. A comparison of salt spray corrosion resistance however does give some indication of the corrosion protection provided by zinc-rich systems, as shown in Table 1.

Table 1. Comparison of Salt Spray Corrosion Resistance in Terms of Percentage of Red Rust Coverage

Material	12 hours	240 hours	500 hours
Zinc-rich paints	0	0	<1
HDG, 1-1/4 oz.	0	25-50	50+
Zn/Fe alloy	0	90+	100
Cold-rolled steel	100	100	100

The adhesion of zinc-rich systems is good when tested by routine bend, impact, or Olsen button deformation, or even when subjected to stamping die operations. No loss in film adhesion is noted even in severely drawn areas limited by the formability of the base steel. Because the coating does not flake or pick off in the stamping dies, scoring or scratching problems are minimized. As a result of the application uniformity obtainable by coil coating of zinc-rich paints, print-through defects do not occur on critical exposed surfaces. Shown in Figure 6 is a typical example of a complex automotive part demonstrating the film integrity even on such severely deformed structures. Similar stamped sections exhibited no red rusting after 240 hours in 5% salt spray tests at the deep drawn corner areas.

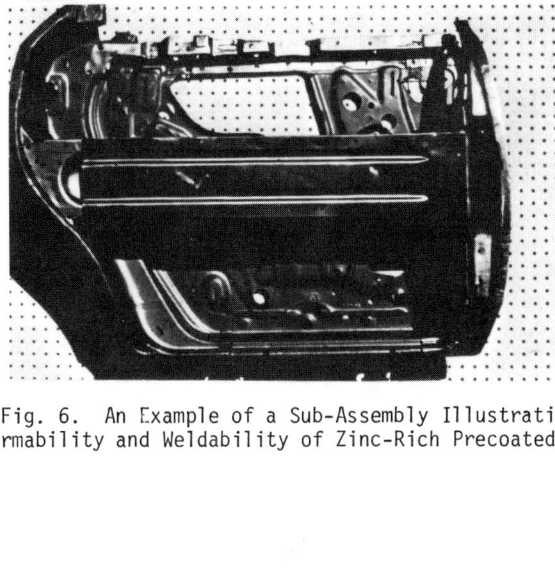

Fig. 6. An Example of a Sub-Assembly Illustrating the Formability and Weldability of Zinc-Rich Precoated Steel

The formed zinc-rich coated parts are resistant to hot alkali spray cleaners and rinsing cycles incorporated into most post-painting processes. Zinc-rich paints are compatible with a variety of finish coat paint systems, providing good adhesion and improved underfilm corrosion resistance. The conductivity of zinc-rich primers allows the deposition of both anodic and cathodic electrodeposited paints now in use throughout the finishing industry. Laboratory evaluations with such duplex coatings show improved performance after salt spray exposure compared to electropaints applied to cleaned and phosphated steel, as shown in Figure 7. The advantage of this duplex system is in providing assurance of corrosion protection to even severely recessed areas of parts if such areas limit adequate electrophoretic deposition.

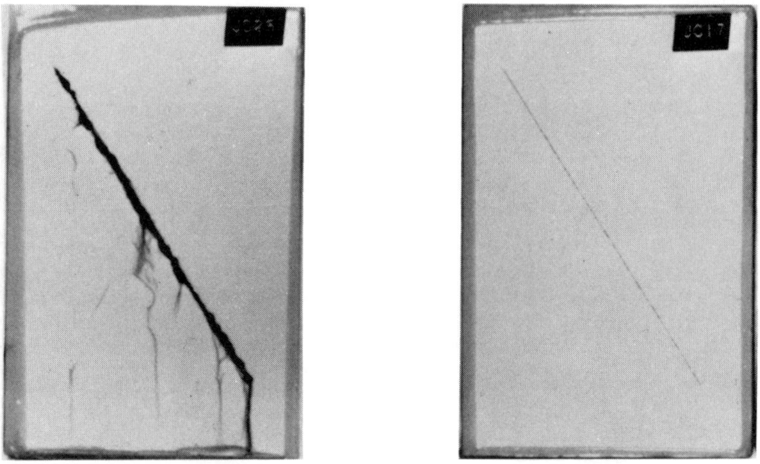

Electropaint over cold-rolled steel

Electropaint over zinc-rich paint

Fig. 7. Salt Spray Corrosion Performance of Electrodeposited Paints

The major producer of a zinc-rich system is Diamond Shamrock Corporation who developed Zincrometal®. Zincrometal® is a two-coat system comprised of a Dacromet® base coat and a Zincromet® top coat. Dacromet® is a proprietary combination of chromic acid, zinc dust and other chemicals which, when cured, provides the conductivity and passivation for the Zincromet® top coat zinc-rich primer. The growth of Zincrometal® precoated steel over the past few years has been remarkable. From its first commercial applications in 1969, Zincrometal® production has reached an annual level of over a million tons.

By far the largest market for zinc-rich precoated steel has been the automotive industry with their concerns of vehicle corrosion from increased road salt usage. Because of the outstanding properties of the zinc-rich systems and the versatility of the coil coating application process, other market areas are being investigated. Different applications may not require the same coating properties deemed necessary for the automotive industry. Zinc-rich systems can be tailored to produce the desired properties for other specific end uses. Automotive manufacturers require a highly critical one-side product with optimum property performance. Other major industrial applications may not require such stringent welding parameters and would desire a two-side coated material as a corrosion resistant primer for exposed decorative finish paints. Such an application could then use a phosphate pretreatment applied in the wet section of the coil line with a one-coat zinc-rich system which would still be weldable and yet provide an excellent corrosion resistant paint base.

Improved zinc-rich paints are required to maintain the production growth of such systems. The coating property requirements, of course, must be met or improved by any new system. But a single-coat zinc-rich paint applied directly to the steel substrate would provide a great advantage by eliminating the present double-cure cycle and thereby reduce gas consumption at the coil line. Low cure temperature zinc-rich systems are a major goal not only for energy considerations and increased coater line speeds, but the reduced peak metal temperature may be required for the new high-strength steels which may be more temperature sensitive than the aluminum-killed drawing quality steels commonly used today. Paint systems may have to incorporate a catalyzed cure mechanism to achieve such low temperature requirements and the versatility of the coil coating lines make this approach feasible. Even the use of different metallic pigments or combinations of pigments should be considered to achieve the basic performance properties.

Zinc-rich paints have been proven to be a corrosion resistant coating system that can be welded, formed, cleaned and topcoated with the versatility and control which makes zinc-rich precoated steel unique to the coatings industry.

REFERENCES

(1) H. H. Uhlig, Corrosion and Corrosion Control, 2d ed., John Wiley & Sons, Inc., New York, 1971.

(2) R. Burns and W. Bradley, Protective Coatings for Metals, Reinhold, New York, 1967.

(3) K. B. Tator, "Zinc-Rich Coatings Forge to Leadership in Corrosion Control," Proceedings of the National Zinc-Rich Coatings Conference,

(3) conducted by the Zinc Institute, Inc. (December 4, 1974).

(4) A. W. Kennedy, "You Can Fool Mother Nature with Zincrometal," Proceedings of the National Zinc-Rich Coatings Conference, conducted by the Zinc Institute, Inc. (December 4, 1974).

(5) C. A. Dahlke, Jr., "Zinc-Rich Primers: Choice and Selection," Materials Protection and Performance, X, No. 9 (September, 1971).

CATHODIC ELECTRODEPOSITION PROVIDES CORROSION PROTECTION

Frederick M. Loop
PPG Industries

Introduction

The development of Cathodic type electrodeposition primers for the automotive industry required new chemistry to produce organic coatings having unique corrosion resistance characteristics. These materials were designed to be applied using the electrodeposition process to assure uniform coating of parts having deep recesses or difficult to coat areas.

It was necessary that the coatings could be applied utilizing existing electrodeposition facilities with modest modifications to minimize equipment costs and to be able to make rapid conversion from Anodic to Cathodic paints feasible. They also had to be compatible with the existing metal pretreatment and paint baking facilities.

The products developed received extensive laboratory and pilot evaluation before being approved for use in production equipment during 1976. The technology was rapidly accepted and most North American automotive electrodeposition facilities are now using Cathodic electrodeposition primers.

Protection of Metal

Metal surfaces can be protected from corrosion through the use of organic primer coatings. These primer coatings are designed to adhere well to the metal surface and provide a barrier film of approximately 12 to 25 microns to prevent corrosion. Specific resin types such as epoxies, epoxy

---

Reprinted with permission of National Association of Corrosion Engineers, Corrosion '80, Paper No. 125, Chicago, March 1980.

esters, alkyds, and others are commonly used to provide these desired characteristics. The resin system used in the Cathodic Electrodeposition primer has these desirable paint characteristics. In addition, corrosion inhibitors are also compounded into the primer composition to increase the protection provided by the primer. These corrosion inhibitors are chromates molybdates or other compounds that have exhibited the ability to improve or extend corrosion protection.

Electrodeposition Process

The method of applying primers using electrodeposition has become quite common in the automotive industry since it was introduced during the 1960's because of its inherent processing advantages. Some of these are:

1. Automation

This method of painting fits well into automotive processing systems either as an in-line processing method or as an off-line system. The coating time is short, 2 to 3 minutes full immersion, for complete body assemblies or parts, and it coats the exterior, interior, and recessed areas in a single dip application. One technician normally operates the processing system.

2. Reliability

Each unit to be coated is fully immersed in the electrodeposition bath. The electrodeposition process using properly formulated materials deposits a uniform coating over all electrically conductive surfaces in the alloted coating time.

3. Ecology

The electrodeposition bath is primarily water. The resin and pigment solids usually consist of not more than 22% of the bath composition. There are also small amounts of solvent present, approximately 5% or less. This composition and the process efficiency are very compatible with the federal and state goals to minimize air pollution.

4. Efficiency

The process results in high utilization of the paint when used with a "closed loop" ultrafiltration system. Properly designed systems have a paint utilization efficiency approaching 100%. Most systems are designed to achieve this high efficiency.

5. Other

There are numerous other characteristics of the electrodeposition process which are advantages, such as non-flammability and freedom from sags or drips.

Cathodic Paint

It was recognized that Cathodic electrodeposition paints would have some inherent advantages in corrosion protection compared with Anodic type paints. These were:

A. Migration Polarity -- Cathodic paints migrate to the metal cathode during the direct current deposition process. Metal ions also tend to migrate to the cathode under the same current flow.

When Anodic paints are electrodeposited, the paint migrates to the metal anode. The anode (the part being coated) initially tends to dissolve under the electrical current flow, but this reaction is smothered by the deposition of the Anodic paint. The tendency for the metal anode to dissolve is also considered to be the first stage of the corrosion process and could initiate corrosion sites on the part being coated.

B. Resin Characteristics -- The Cathodic resin consists of alkaline polymers solubilized in water, using water soluble acids to produce a nearly neutral paint (pH 6.0 to 7.0).

In the electrodeposition process, the alkaline polymers migrate to the cathode coating the part with the alkaline resin while the water soluble acid migrates to the anode. Thus, the part is coated with the alkaline polymers which inherently are rust inhibitors.

Anionic electrodeposition resins consist of acidic polymers, solubilized in water, using water soluble alkalies to produce an alkaline paint (pH 7.0 plus). In the electrodeposition process the acid polymers migrate to the anode, coating the part with the acidic resin while the alkali migrates to the cathode. Acid type polymers are not rust inhibitors. This reaction is illustrated in Figure 1.

Corrosion Testing

The most commonly used laboratory test for corrosion resistance is the salt fog exposure ASTM B 177. This test is easily performed, readily controlled, and produces results in relatively short periods. The laboratory results correlate fairly well with corrosive environments to which automotive units are exposed in actual usage, particularly in the snow areas where salt is used on streets and highways. Test results of corrosion resistance tests for Cathodic and Anodic primers on various substrates are reported in Table I.

TABLE I

Corrosion Resistance Tests
ASTM B-117

Paint	Substrate	Pretreatment	Exposure Time (Hrs)	Scribe Creep (Mm.)
Cathodic	Steel	Zinc Phosphate	336	0.5
Anodic	Steel	Zinc Phosphate	336	1.0
Cathodic	Steel	NONE	336	2
Anodic	Steel	NONE	96	5
Cathodic	Galvanize	Zinc Phosphate	336	1
Anodic	Galvanize	Zinc Phosphate	336	0.5
Cathodic	Aluminum		336	0.5
Anodic	Aluminum		336	0.5

Bake 30 min. @ 350°F.
Film Thickness 15 to 20 microns
Zinc Phosphate - EP-1
Scribed Panels - Blown with 90 lb. air pressure

Because of the slow rate of corrosion failure on Cathodic coated panels, some laboratories are extending the salt fog exposure period to 672 hours or longer to produce significant scribe creepage.

For electrodeposition primers, another laboratory test is used to determine the ability of the primer to protect deeply recessed areas. This test is known as the "throw-power" test.

The test unit consists of a rectangular box formed from two 10.1 cm. by 45.7 steel panels which are spaced apart 9.5 mm. with two rubber side gaskets. The open ends of the box have a 9.5 mm. by 10 cm. opening as noted in Figure 2.

The box is coated by immersing the unit into a coating tank to a depth of about 43 cm. The test box is coated at a voltage and time sufficient to produce a normal coating on the exterior of the box. When disassembled after being coated, the test unit produces two test panels which have a coating thickness on the inner side varying from approximately 16 microns tapering down to no coating. The distance from the bottom opening to the point where no paint is deposited is measured and rated as the "throw power" of the paint.

The test panels can be baked and exposed to salt fog to evaluate the corrosion resistance of the primer over various surfaces at very thin films. Results of "throw power" tests comparing Cathodic paint to Anodic paints are noted in Table II. 28 cm of "throw-power" is considered adequate to coat the interior sections of most automotive units.

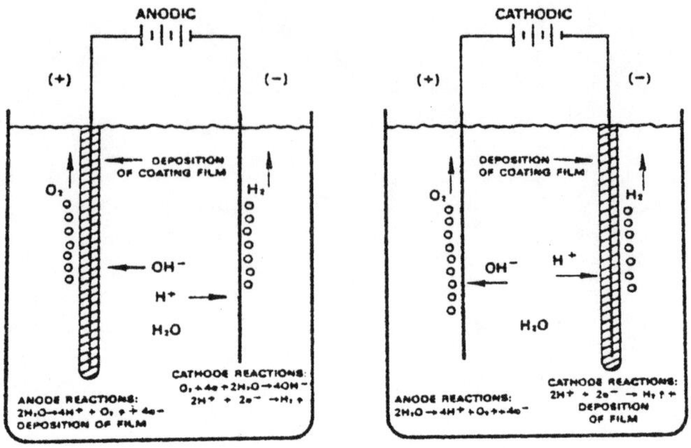

FIG. 1 - Electrodeposition

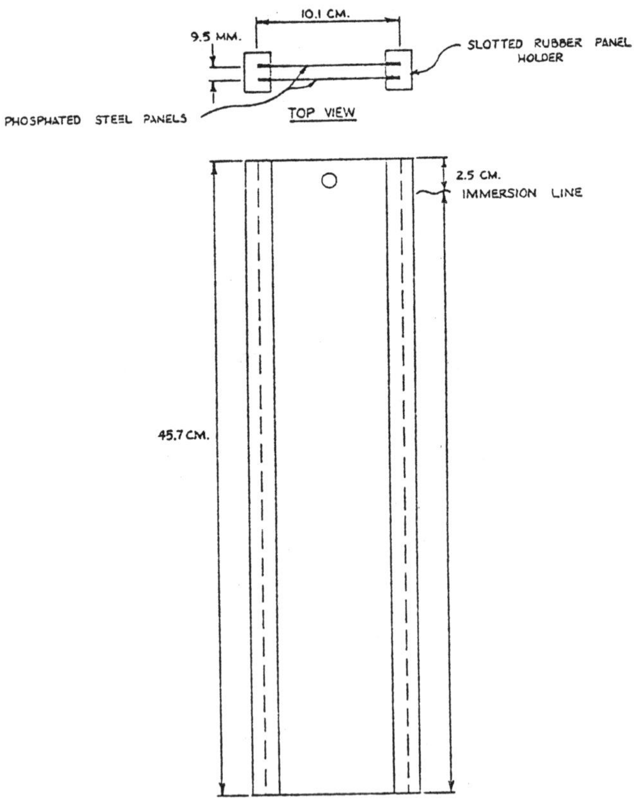

FIG. 2 - Two panel throw box

TABLE II

Throw-Power Test

Paint	Coating Time (Min.)	Test Voltage	Throw-Power (Cm)
Cathodic	2	300	31
Anodic Polyester (I)	2	400	31
Anodic Polyester (II)	2	225	23

EP-1 Zinc phosphated steel panels
Bake 30 min. @350°F.

Results of "throw-power" panels, comparing an Anodic and Cathodic paint on zinc phosphated steel steel panels before and after salt fog exposure are shown in Figure 3A and B.

There are also numerous other laboratory test methods used to evaluate corrosion on panels. These are usually a combination of salt fog ASTM B 117 with some other environmental exposure such as freezing or high humidity and temperature. The Cathodic type primer produces results consistently better than Anionic or conventional primers in these tests.

Pilot test cars were painted in large Cathodic dip test tanks and the cars were subjected to a test track corrosion exposure cycle sequence to prove out performance on units. A test sequence cycle consists of driving the test car through a series of salt water showers and salt water splash troughs, plus driving at various speeds over a variety of roads. Included in each test cycle is a high humidity exposure period at high temperatures, e.g. 49°C.

This test method is considered the most accurate projection of performance expected in actual severe field conditions. Units are usually tested for 40 to 60 cycles, but some Cathodic cars have been subjected to more than 80 cycles. The Cathodic coated cars exhibited considerably less visible corrosion and fewer perforations than non-Cathodic coated cars after the most severe exposure tests.

FIG. 3A - Cathodic

FIG. 3B - Anodic

Paint Characteristics

The applied Cathodic primers also had to have other necessary characteristics to meet the stringent automotive requirements for coatings used on the exterior of the car. Some of these were:

TABLE III

	CHARACTERISTIC	TEST	RATING
1.	Chip Resistance	SAE J-400 Gravelometer	Very Good
2.	Humidity Resistance	ASTM D 2247	Pass 240 hours
3.	Smoothness	Visual	Very Good
4.	Hardness	Pencil	2H plus
5.	Florida Exposure (with topcoat)	2 Years	Pass

The paint material had to have characteristics satisfactory for its use in production facilites. Typical characteristics of the paint as applied are:

TABLE IV

PROPERTY	PARAMETER
Total Solids	20-22%
pH	6.0-7.0
Voltage (Application)	250-400
Temperature	16-27°C
Film Thickness	15-22 Microns
Bake Schedule	30 min. @ 177°C
Stability	3 months minimum

Equipment

The Cathodic paint uses equipment which is common to many Anionic type electrodeposition systems. Some of the specific characteristics of the equipment are noted in Table V.

TABLE V

	CHARACTERISTIC	TYPE
1.	Materials of Construction	Mild Steel
2.	Dip Tank	Lined
3.	Polarity	Part is the cathode and grounded
4.	Power Supply	Up to 500 volts D.C.
5.	Ultrafiltration	"Closed loop" design
6.	Solubilizer Control	Membrane System
7.	Waste Treatment	Cathodic materials

Figure 4 is a schematic diagram of a Cathodic type system.

Current Status

In North America, more than 95% of automotive electrodeposition systems are using Cathodic primer. The change from Anodic to Cathodic type materials has occurred very rapidly. This change was made as a part of the automotive industries effort to provide units having significantly improved corrosion resistance. All planned or proposed new electrodeposition systems are scheduled to use Cathodic primers.

Evaluation of units in the field has indicated successful corrosion protection, although the exposure period must be extended for several years to indicate the degree of improvement.

Ecology compliance programs include electrodeposition as a basic system to provide minimal air pollution. The electrodeposition systems have high transfer efficiency (approaching 100%) and have low solvent content. These two characteristics are critical in meeting the latest federal and state air pollution control proposals.

Future

The technology of Cathodic resin systems is new, but the initial paint materials achieved a high level of corrosion resistance compared to the standard solvent borne

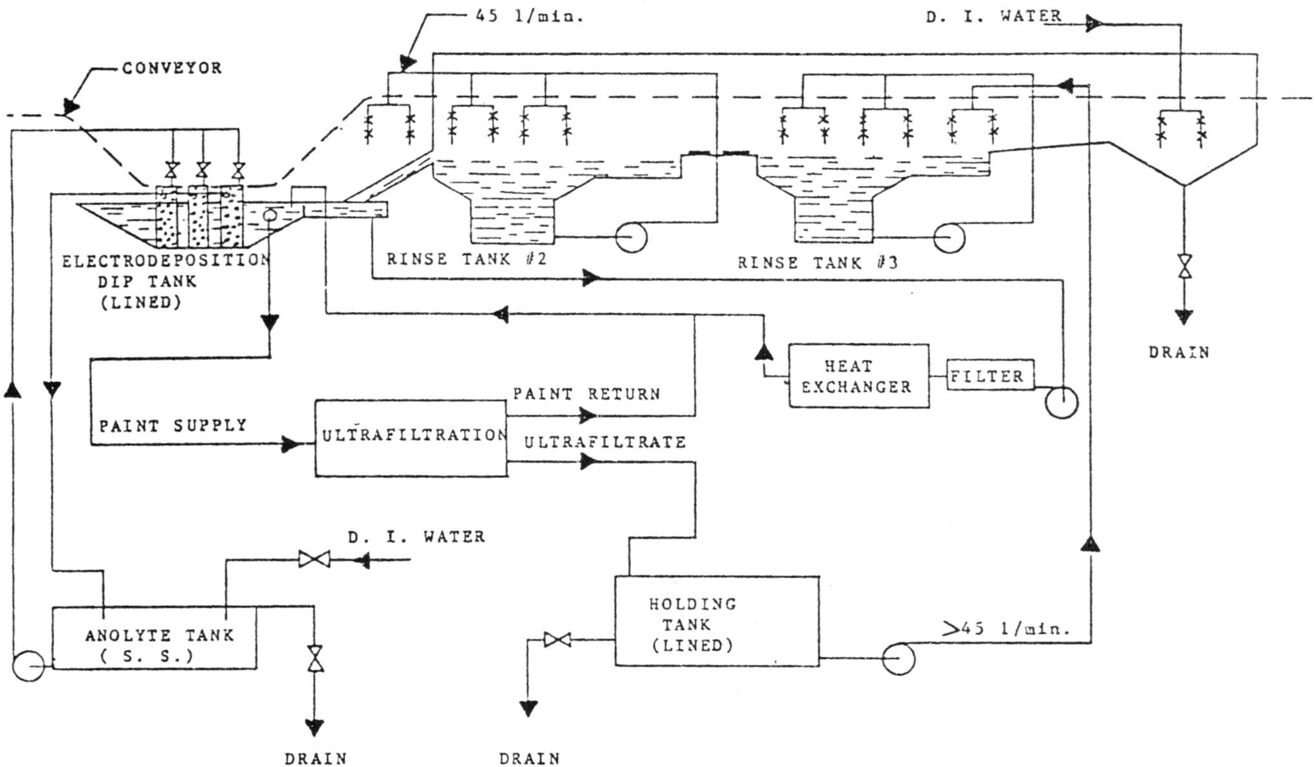

FIG. 4 - Cathodic electrodeposition system

and Anodic materials in use. This quality improvement was achieved over various substrates such as cold rolled steel, phosphated steel, galvanize, aluminum and others. Further improvements in corrosion resistance on steel are possible through specific metal pretreatment systems designed for use with Cathodic paint.

It is possible to modify the Cathodic paints as necessary to provide characteristics such as chip resistance, higher film build, color variation, and other properties.

Specific materials are being developed to provide lower baking temperatures for use with plastic parts or for further energy reduction.

It is expected that Cathodic materials can be developed to meet any specific paint characteristic and maintain superior corrosion resistance.

CORROSION PROTECTION — II:

MATERIALS AND DESIGN IN HIGHWAY AND OFF-HIGHWAY VEHICLES

UNDERVEHICLE CORROSION RESISTANCE OF 55% Al-Zn-COATED STEEL SHEET

L. Allegra and H. E. Townsend

Corrosion and Coatings Research Section,
Research Department, Bethlehem Steel Corporation,
Bethlehem, PA 18016.

Al-Zn-coated steel sheet was developed by Bethlehem Steel's Research Department and is commercially available in the United States under the trade name Galvalume. The coating, applied by a continuous hot-dip process, is an alloy of 55% Al, 43.4% Zn and 1.6% Si (nominal weight percentages). This composition was selected on the basis of corrosion tests of a series of Al-Zn alloy coatings in rural, industrial and marine atmospheres. The test results showed that the 55% Al composition exhibits an optimum combination of the durability of aluminum coatings and the sacrificial characteristics of zinc coatings [1]. Adding the coatings of both sides gives a nominal coating weight of 0.5 oz/sq ft (153 g/sq m) corresponding to a total coating thickness of 1.6 mils (40 micrometers). The product is readily formable and is fastened or joined by any of the conventional techniques used for coated steels. Typical mechanical properties for commercial quality sheet are:

Yield Strength	38,000-53,000 lb/sq in. (262-365 MPa)
Tensile Strength	50,000-65,000 lb/sq in. (345-448 MPa)
Total Elongation	20 to 36%

Atmospheric tests lasting up to 13 years have amply demonstrated the excellent durability and long-term corrosion protection provided to steel sheet by the 55% Al-Zn coating [2]. Comparison of the corrosion rates for a wide range of environments has shown that: (a) the 55% Al-Zn is at least two to four times more durable than zinc coatings of equal thickness, and (b) after about 10 years or more exposure in industrial and rural environments the corrosion rates of 55% Al-Zn coatings are indistinguishable from those of aluminum coatings.

Atmospheric testing also demonstrated the ability of the 55% Al-Zn coating to provide cut-edge protection to steel sheet. Both the 55% Al-Zn and galvanized coatings provide the galvanic protection characteristic of zinc. However, the greater durability of the 55% Al-Zn coating assures edge protection for much longer periods of time. Compared to aluminum coatings, which provide sacrificial protection only in marine environments, the alloy coating provides galvanic protection at cut edges in all environments, including industrial and rural. Owing to its durability and other favorable properties, unpainted 55% Al-Zn-coated steel sheet has become one of the most widely used materials for preengineered metal building roofing in the U.S., where it is replacing both aluminum-coated and prepainted galvanized steel sheet [3].

Benefits realized in the case of metal roofing may also be possible in automotive applications. As indicated in a recent review [4], road salt is a principal cause of automobile corrosion. When tested in salt spray with cut edges exposed, 55% Al-Zn-coated steel is at least two times more corrosion resistant than aluminum-coated Type 1 or G90 galvanized steel [5]. Since galvanized steel is known to perform well in the autobody environment [6,7,8], there is ample reason to anticipate that Al-Zn-coated sheet will exhibit excellent durability in this application. The purpose of this paper is to report the first results of tests comparing galvanized and Al-Zn-coated steel in the actual undercar environment.

G90 galvanized and 55% Al-Zn-coated steel sheets (Table 1) were sheared into 2x5-inch (5.08x12.7-cm) coupons, impressed with 0.250-inch (6.4-mm) Olsen cups, weighed, and bolted together in nested pairs with a 10-mil (0.3-mm) gap between coupons. Nested pairs were then bolted to stainless steel racks, which were subsequently mounted beneath passenger cars on the rear-axle control arms (Figure 1) according to the test procedure described in detail by Neville [9].

After removal from service, coupon pairs were disassembled, soft-bristle brushed in water, cleaned in chromic acid according to the ASTM G1-72 procedure for zinc, and weighed to determine corrosion losses. Weight losses were converted to thickness losses using the formula:

$$t = \frac{1}{10,000} \quad \frac{W}{AD}$$

TABLE 1. MATERIALS IN THE UNDERCAR PANEL TEST

Coupon Pair	G90 Galvanized Steel: Sheet Thickness inch (mm)	G90 Galvanized Steel: Coating Thickness* mil (μm)	55% Al-Zn-Coated Steel: Sheet Thickness inch (mm)	55% Al-Zn-Coated Steel: Coating Thickness* mil (μm)
A	0.043 (1.09)	0.8 (20)	0.033 (0.84)	1.2 (30)
B	0.043 (1.09)	0.8 (20)	0.033 (0.84)	1.2 (30)
C	0.043 (1.09)	0.8 (20)	0.033 (0.84)	1.0 (25)
D	0.022 (0.56)	1.0 (25)	0.033 (0.84)	1.0 (25)
E	0.022 (0.56)	0.8 (20)	0.033 (0.84)	0.9 (23)
F	0.035 (0.89)	1.0 (25)	0.031 (0.79)	1.0 (25)

* Average of four coupon faces determined nondestructively using the magnetic thickness technique (ASTM E376-69).

where

t = average thickness loss per surface, μm
W = weight change per coupon pair, g
A = total surface area per coupon pair, 258.1 sq cm
D = density of coating metal,
7.13 g/cu cm for galvanized and
5.25 g/cu cm for the zinc-rich portion of the 55% Al-Zn coating.

The density of a 22% Al-Zn alloy was used as an approximation to the density of the zinc-rich constituent that was observed to corrode preferentially [1] from the 55% Al-Zn coating. This approximation was confirmed by determinations of actual compositions of corroded constituents on several of the undercar coupons.

FIGURE 1. Undercar Corrosion Coupon Rack Mounted on Rear-axle Control Arm of a Passenger Car

The undercar panels were exposed for a wide range of service times and mileages in the Bethlehem, Pa. area, but all racks were driven at least one full winter (Figure 2).

No rust was observed on any of the panels of either material. Sheared edges were free of rust, indicating excellent galvanic protection by the galvanized and 55% Al-Zn coatings. Corrosion rates calculated from the weight loss data (Table 2) vary considerably, indicating that the particular driving environment is an important corrosion factor.

Corrosion rates for galvanized were higher in every case, averaging 1.81 micrometers/year compared to 0.45 micrometers/year for the 55% Al-Zn coating. This difference is significant at the 99% confidence level as determined by the Students-t comparison. At these rates a 20 micrometer galvanized coating would be expected to protect the steel base for about 11 years in the Bethlehem environment. Other areas of the country may be more corrosive. However, since the superiority of the 55% Al-Zn coating is maintained over a three-fold range of severity, it is expected that the alloy-coated sheet would be more durable than galvanized over a wide range of geographical areas.

Serial removals from the same vehicle, Pairs D and E, showed that (a) corrosion rates for both materials decrease with time, (b) the ratio of galvanized to 55% Al-Zn remains nearly constant at about two to one, and (c) corrosion of the 55% Al-Zn coating is less than galvanized even though the exposed steel edge is about 50% greater for the alloy-coated panels.

In summary, the three-year tests on road-driven cars showed that compared with galvanized the 55% Al-Zn coated steel sheet is two to eight times more resistant to under-vehicle corrosion.

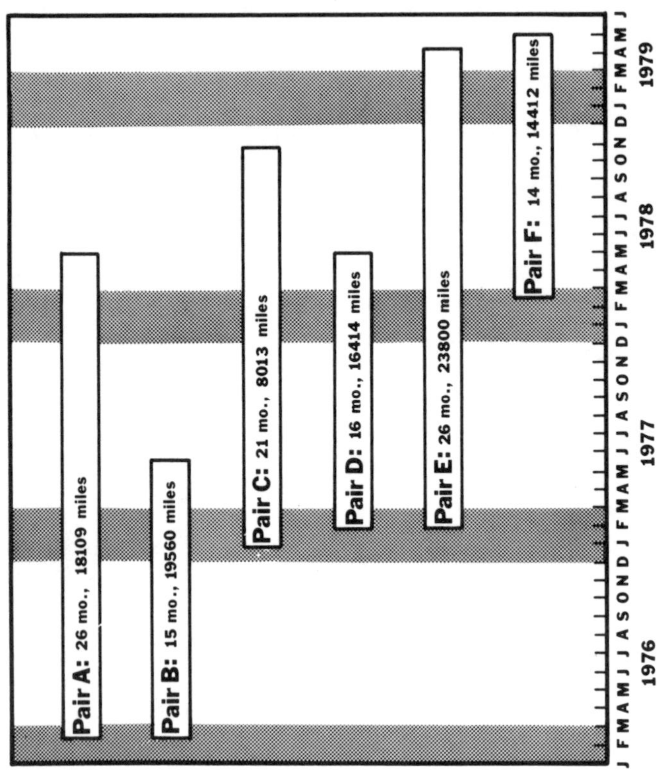

FIGURE 2. Service Schedule for Undercar Racks. Shaded Areas are Winter Periods.

TABLE 2. WEIGHT LOSSES AND CORROSION RATES FOR UNDERCAR PANELS

Coupon Pair	Weight Loss mg		Corrosion Rate μm/yr		Relative Severity**	Ratio of Corrosion Rates: Galvanized to 55% Al-Zn-Coated
	Galvanized	55% Al-Zn-Coated	Galvanized	55% Al-Zn-Coated		
A	1327.3	137.0	3.33	0.47	3.0	7.1
B	359.1	71.6	1.56	0.42	1.4	3.7
C	495.2	48.6	1.54	0.21	1.4	7.3
D*	344.4	122.2	1.40	0.68	1.3	2.1
E*	455.8	189.1	1.10	0.65	1.0	1.7
F	407.7	37.9	1.90	0.24	1.7	7.9
		Average	1.81	0.45		

* Serial removals from the same vehicle.

** Compared to galvanized corrosion rate of 1.10 μm/yr (Pair E).

ACKNOWLEDGMENTS

Special thanks to: F. R. Dold for installation and evaluation of undercar racks, those Bethlehem employees who volunteered to have test racks attached to their cars, and B. S. Mikofsky for his editorial assistance.

REFERENCES

[1] Zoccola, J. C., Townsend, H. E., Borzillo, A. R., and Horton, J. B., "Atmospheric Corrosion Behavior of Aluminum-Zinc Alloy-Coated Steel," Atmospheric Factors Affecting the Corrosion of Engineering Metals, ASTM STP 646, S. K. Coburn, Ed., American Society for Testing and Materials, 1978, pp. 165-184.

[2] Townsend, H. E., and Zoccola, J. C., Materials Performance, Vol. 18, No. 10, p. 13 (1979).

[3] Horton, J. B., Borzillo, A. R., Kuhn, N., and Harvey, G. J., "Galvalume and Zincalume -- Sheet Steel Coated with 55% Al-Zn Alloy," paper presented at the Twelfth International Conference on Hot-Dip Galvanizing, Paris, May 1979.

[4] Cracking Down on Corrosion, AISI, Washington (1978), 15 pp.

[5] Allegra, L., and Zoccola, J. C., Materials Performance, Vol. 18, No. 5, p. 18 (1979).

[6] Pearlstein, F., and Gallaccio, A., Journal of Testing and Evaluation, Vol. 3, No. 6, p. 414 (1975).

[7] Burst, H., Grahle, M., Grashoff, J., and Schneider, C., Metal Progress, Vol. 111, No. 6, p. 38 (1977).

[8] Russell, R. J., and Kiuchi, J. T., Materials Performance, Vol. 18, No. 3, p. 16 (1979).

[9] Neville, R. J., SAE Paper No. 770364, presented at the Annual Congress and Exposition, Detroit (1977).

A GALVANIZED-GALVANNEALED SHEET STEEL FOR AUTOMOTIVE BODY PANELS

W. C. Sievert
L. L. Franks
Inland Steel Company

INTRODUCTION

Inland Steel has developed a hot-dip one-side galvanized, one-side galvannealed sheet steel product, PAINT-TITE B,* which possesses inherent characteristics and properties desirable for exposed automotive body panel applications. In this regard, PAINT-TITE B furnishes corrosion resistance from salt environments by a G.90 galvanized coating on the surface for interior exposure to retard inside-out perforation, and by the good underfilm corrosion resistance of a zinc-iron alloy coating on the exterior surface. This presentation describes the production process as related to the appearance and forming, corrosion, painting, and welding performance of PAINT-TITE B as a two-side coated steel product suitable for highly visible body panels.

GALVANIZING-GALVANNEALING TECHNOLOGY

The specific goals with regard to control of the iron diffusion kinetics involved in hot-dip galvanizing are diametrically opposed to those in galvannealing. In the case of hot-dip galvanized coatings, the addition of aluminum to the molten bath is for the expressed purpose of retarding the growth of interface zinc-iron alloy. In addition, thermal input to the steel is closely regulated in order not to overcome this retardation. However, in the case of galvannealing, the goal is to overcome the retardation effect of aluminum by raising the thermal input above a specific level and cause iron diffusion to proceed rapidly into the zinc coating to form a zinc-iron alloy coating.

*PAINT-TITE B is a registered trade name of Inland Steel Company.

Therefore, for the production of PAINT-TITE B, precise control over the thermal input must be exercised in order to form a completely zinc-iron alloy on one surface, while minimizing the zinc-iron alloy growth on the other surface. Figure 1 shows a schematic cross-section diagram of the coated steel with a G.90 galvanized coating on one surface and an A.01 zinc-iron alloy on the other.

Figure 2 shows the product at the galvanizing line recoil unit, exhibiting the characteristically bright hot-dip galvanized coating on one surface and the matte gray galvannealed coating on the other.

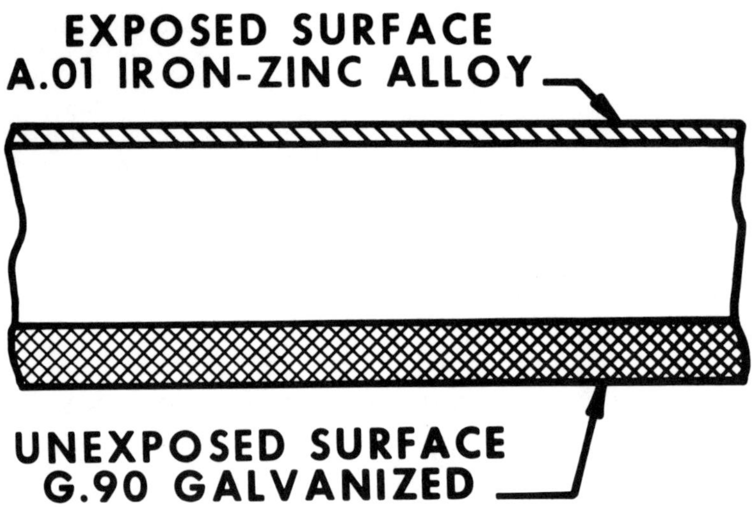

Fig. 1. Schematic Cross-Sectional Diagram of PAINT-TITE B

Fig. 2. PAINT-TITE B at the Galvanizing Line Recoil Unit

Though careful control of the zinc-iron alloy growth is exercised, in actual practice an appreciable amount of zinc-iron alloy growth does occur at the interface of the G.90 coating. Since rapid heat transfer occurs through the strip from one side to the other, the thermal input is regulated to just complete the zinc-iron alloy growth through the 3.8μ m (0.15 mil) zinc coating, while restricting the growth on the other side.

The iron diffusion mode into the thin, 3.8 μ m (0.15 mil) zinc coating is illustrated by three SEM photomicrographs shown in Figure 3. The change from a cast zinc coating to a zinc-iron alloy coating is accompanied by both a change in the macro-appearance from a bright, shiny coating to a matte gray, and by a change in the microstructure from the smooth finish of zinc to the porous structure of zinc-iron alloy.

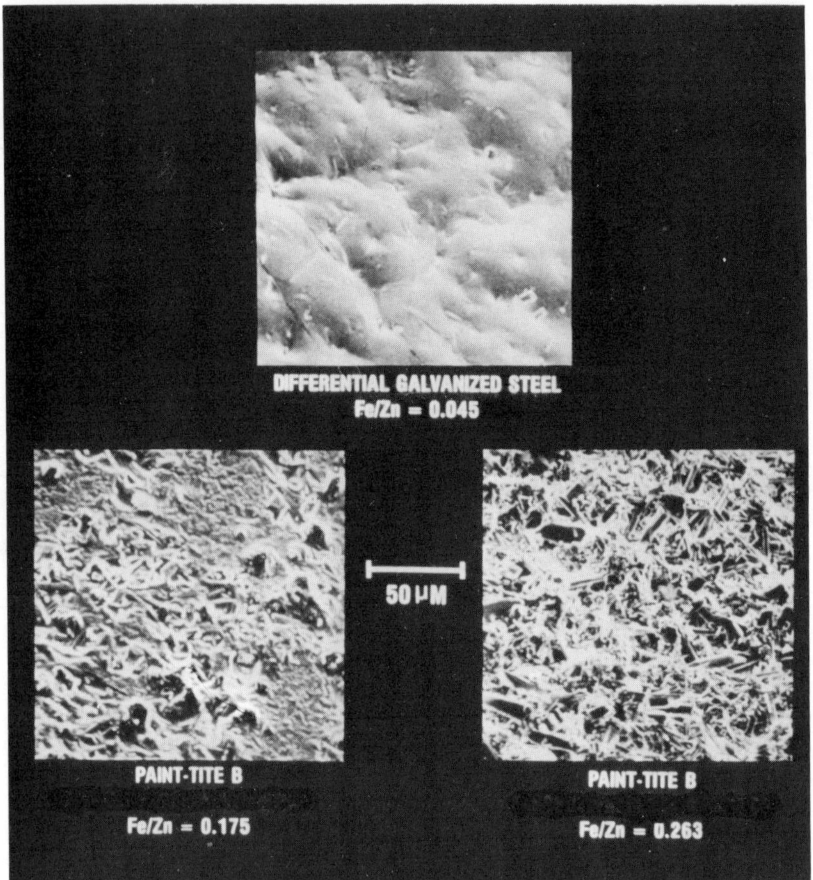

Fig. 3. SEM Photomicrographs of the Galvannealing Process Sequence on A.01 Coating

By restricting the growth at the coating-steel base interface, good formability properties are retained by the relatively thick, ∿19 μm (0.75 mil) G.90 coating.

MICROSTRUCTURAL FORMABILITY RELATIONSHIP

Figure 4 shows SEM photomicrographs of a typical thin, A.01 zinc-iron alloy which shows good formability by various mechanical deformation methods. Figure 5 shows cross-section photomicrographs of two different G.90 galvanized coatings indicating different degrees of zinc-iron alloy growth, while still maintaining good forming properties. The good ductility of these coatings may be explained by the suppression of the continuous layer of hard zinc-iron intermetallic compound at the interface. Microhardness determinations indicate the presence of either a relatively soft alloy layer at the interface or the presence of a hard intermetallic compound in a zinc matrix. Table I shows that the microhardness values obtained were significantly lower than the published figures for the zeta ($FeZn_{13}$) and the delta ($FeZn_7$) phases. Since microprobe analysis indicates the presence of $FeZn_7$, it is reasonable to conclude that this alloy phase has grown discontinuously within the eta (Zn) phase.

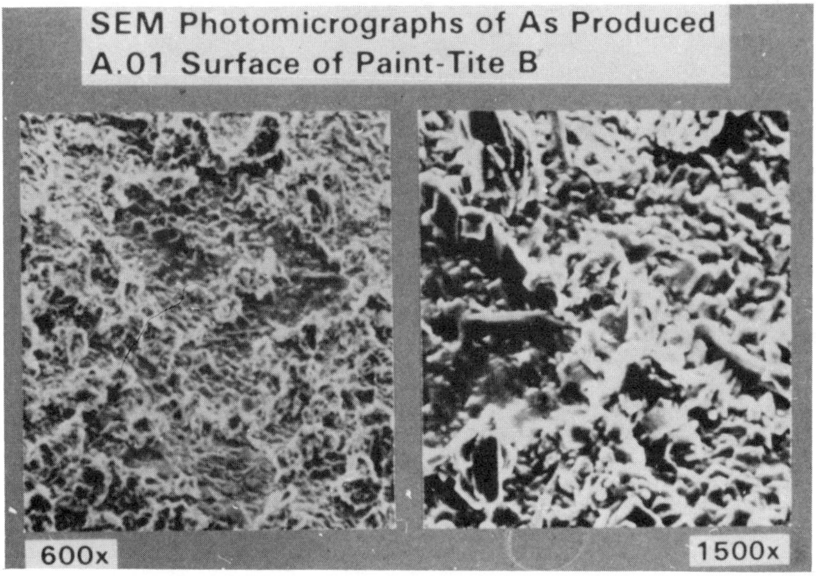

Fig. 4. SEM Photomicrograph of Zinc-Iron Alloy A.01 Coating

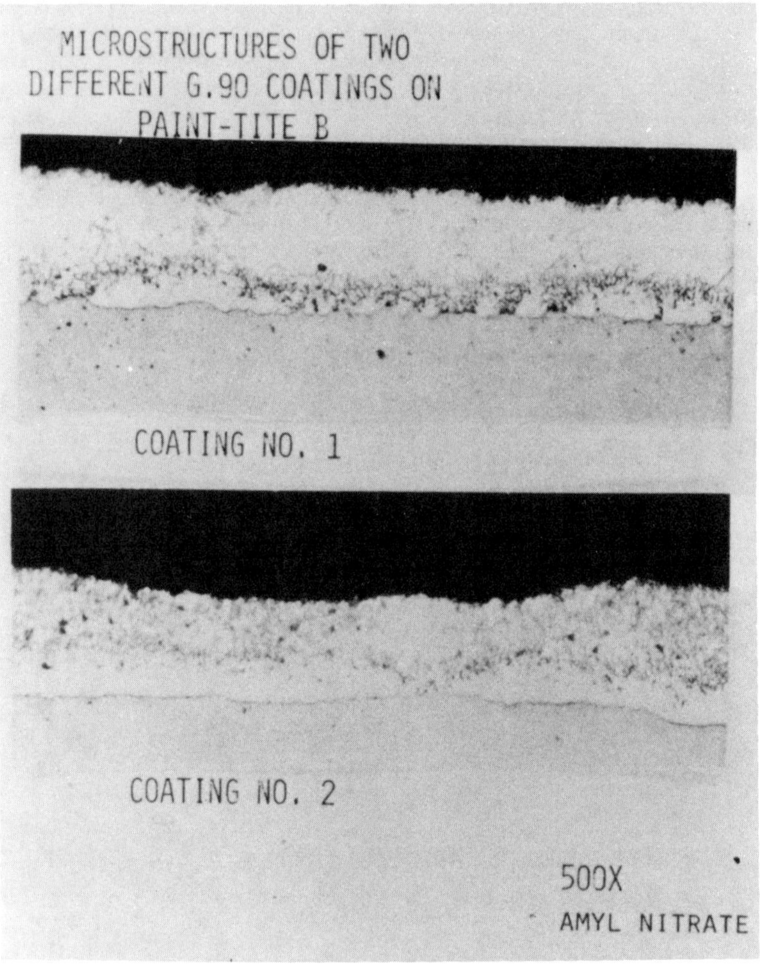

Fig. 5. Cross-Section Photomicrographs of Two G.90 Coatings

Table I. Microhardness Determinations for the Interface Alloy on PAINT-TITE B G.90 Coatings

Test Area or Phases	Microprobe Analysis % Fe by Wt	Microhardness KHN
Sample No. 1 Interface Alloy	8-10	124
Sample No. 2 Interface Alloy	9-11	188
$Delta_1$ ($FeZn_7$)	7-11.5*	376-488*
Zeta ($FeZn_{13}$)	6-6.2*	250-300*

*Published Values

Figure 6 shows 180° O-T bend tests on PAINT-TITE B for the A.01 and G.90 coatings. Only light crazing can be noted, as highlighted by this close-up photograph, with no flaking or powdering of the coating. This good formability exhibited by both coatings is typical even for the relatively large zinc-iron alloy growth in the G.90 coating microstructures described above. Satisfactory stamping plant performance for a wide variety of automotive panels, such as doors, fenders, cowls, and lift-gates, has verified the desirable coating formability characteristics of PAINT-TITE B.

Fig. 6. 180° O-T Bends on PAINT-TITE B

Base metal formability is dependent upon the steel base chemistry and the degree of heat treatment and temper mill extension. Table II shows typical mechanical properties for drawing quality and commercial quality PAINT-TITE B produced by the continuous heat treatment furnace of Inland's No. 5 Galvanizing Line. The drawing quality steel normally includes a post anneal step prior to temper rolling to impart added ductility.

Table II. PAINT-TITE B Typical Mechanical Properties as Temper Rolled

	Yield Strength		Tensile Strength		% Total	Rockwell B
	MPa	(ksi)	MPa	(ksi)	Elongation	Hardness
Drawing Quality Special Killed	227	(33)	345	(50)	39	48
Drawing Quality Rimmed	248	(36)	338	(49)	37	49
Commercial Quality	262	(38)	365	(53)	32	59

PROFILOMETRY-APPEARANCE RELATIONSHIPS

A smooth, defect-free appearance is essential for sheet steel application for exposed automotive panels. In this regard, PAINT-TITE B, because of the surface properties imparted by the thermal process, has ideal characteristics.

The zinc-iron alloy A.01 coating, because of the iron diffusion process, is smooth and spangle free. The thermal process produces a flat profiled G.90 coating with the spangle minimized by the nucleating effect of the partial alloy growth, and tends to smooth out the ripples which are normally imparted in the free-zinc portion by the high velocity gas impingement coating thickness control mechanism. The surface profilometry determinations shown in Table III for temper rolled PAINT-TITE B indicate a uniform profile which is consistent in both the A.01 and G.90 coatings and conforms to the automotive specification. This uniform profilometry translates into a smooth drawn surface, as shown by the 76 mm (3 in) bulge test shown in Figure 7. Further evidence of this desirable characteristic is shown by the smooth finish on the automotive panel shown in Figure 8.

Table III. Surface Profilometry Measurements on Temper Rolled PAINT-TITE B

Surface	Average Roughness R_a μ m(μ in)	Average Measured Peak Count P_k mm(in)	Minimum Peak Count Required
A.01 Coating			
Transverse	1.1 (43)	5.6 (141)	124
Longitudinal	1.1 (43)	5.4 (136)	124
G.90 Coating			
Transverse	1.0 (41)	4.8 (123)	113
Longitudinal	1.0 (40)	4.6 (117)	108

R_a must be within range of 0.5-1.3 μ m (20-50μ in)
P_k must exceed the minimum peak count as calculated by
$5.3\ R_a - 104$
(For Automotive Specification)

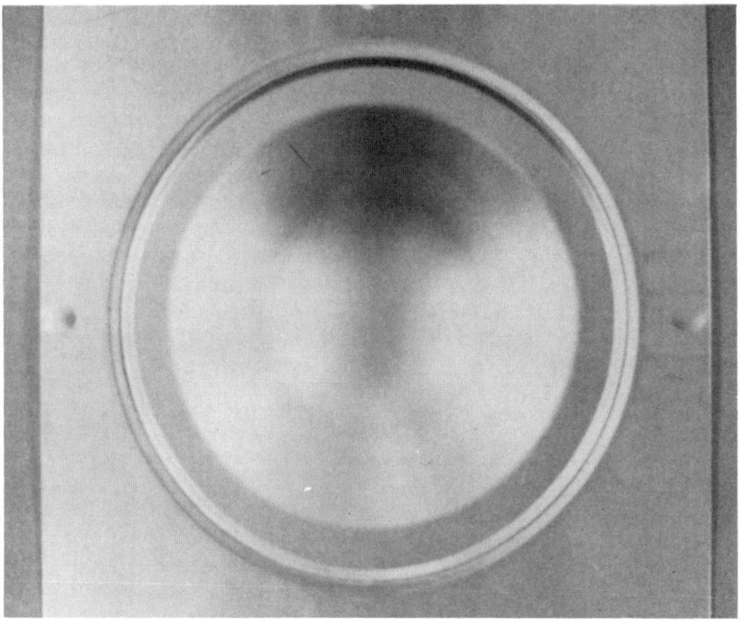

Fig. 7. 76 mm (3 in) Bulge Test

Fig. 8. Door Panel Stamped from PAINT-TITE B

CORROSION RESISTANCE AND PAINTABILITY

The expressed purpose of utilizing a one-side zinc coated, one-side zinc-iron alloy coated sheet steel is to provide inside-out corrosion resistance from the splashed and entrapped road de-icing salts which cause perforation through the base metal, and to provide underfilm protection and a degree of cathodic base metal protection on the exterior surface. The degree of protection furnished by the G.90 coating of PAINT-TITE B is illustrated in Figure 9, which shows 5% salt spray data compared to conventional hot-dip galvanized coatings. Though partial alloying does occur in the G.90 coating, these test results indicate that sufficient "free" zinc combined with the zinc-iron alloy retards red rust to a similar extent as conventional hot-dip galvanized G.90 coatings.

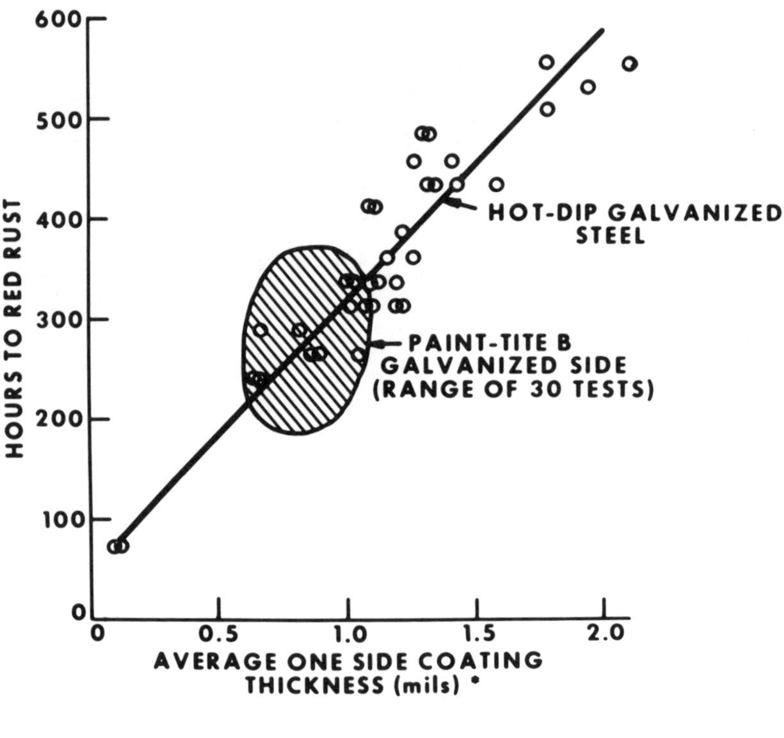

Fig. 9. 5% Salt Spray Exposure Values for PAINT-TITE B

The underfilm corrosion and cathodic protection properties of the zinc-iron alloy coating are exhibited by the comparative painted cold rolled steel and PAINT-TITE B panels shown in Figure 10. These test panels were pretreated with zinc-phosphate and painted with an automotive system. Though the zinc-phosphate effectively retarded extensive creepage, a profusion of red rust exudes from the unprotected scratch through the paint film on the cold rolled steel. The electro-negative potential of the zinc-iron alloy, however, protects the steel base until remedial touch-up painting can be accomplished. Figure 11 shows two doors stamped and painted by an automotive company, one from conventional cold rolled steel and the other from PAINT-TITE B. After 1000 hours exposure to 5% salt spray, a significant amount of red rust bleeding occurred from a scratch through the paint on the cold rolled steel substrate.

Fig. 10. 5% Salt Spray Exposure of Painted Cold Rolled Steel and PAINT-TITE B

Fig. 11. Painted Doors Exposed 1000 Hours in
5% Salt Spray - PAINT-TITE B vs Cold Rolled Steel

WELDABILITY

Zinc-iron alloy coatings provide better spot welding performance characteristics than zinc coatings. This desirable characteristic of the A.01 zinc-iron alloy coating is illustrated in Figure 12 by the greater number of spot welds achieved before electrode dressing was required than on a zinc coating of equivalent thickness. This performance was achieved by welding with the electrode to coupon orientation shown in Figure 13. The essential compositional differences between the differentially coated hot-dip galvanized steel used in this test and PAINT-TITE B are shown in Table IV. The iron-zinc ratio values shown reflect the degree of iron diffusion that occurred to produce the zinc-iron alloy coating that contacts the bottom electrode.

Table IV. PAINT-TITE B A.01 and G.01
Coating Characterization

Material	Heavy Side Coating Weight		Light Side Coating Weight		Light Side	Light Side Relative Depth to Fe Detection**
	g/m^2	(oz/ft^2)	g/m^2	(oz/ft^2)	I_{Fe}/I_{Zn}*	
PAINT-TITE B	144.5	(0.47)	18.0	(0.059)	0.263	0.04
Differentially Galvanized Steel	137.0	(0.45)	16.8	(0.055)	0.045	1.00

*Characteristic K x-ray intensity ratio at the light coating surface.

**Ion scattering spectrometer (ISS) sputtering time before detection of Fe, values shown relative to differentially galvanized steel light side coating value standardized to be 1.00.

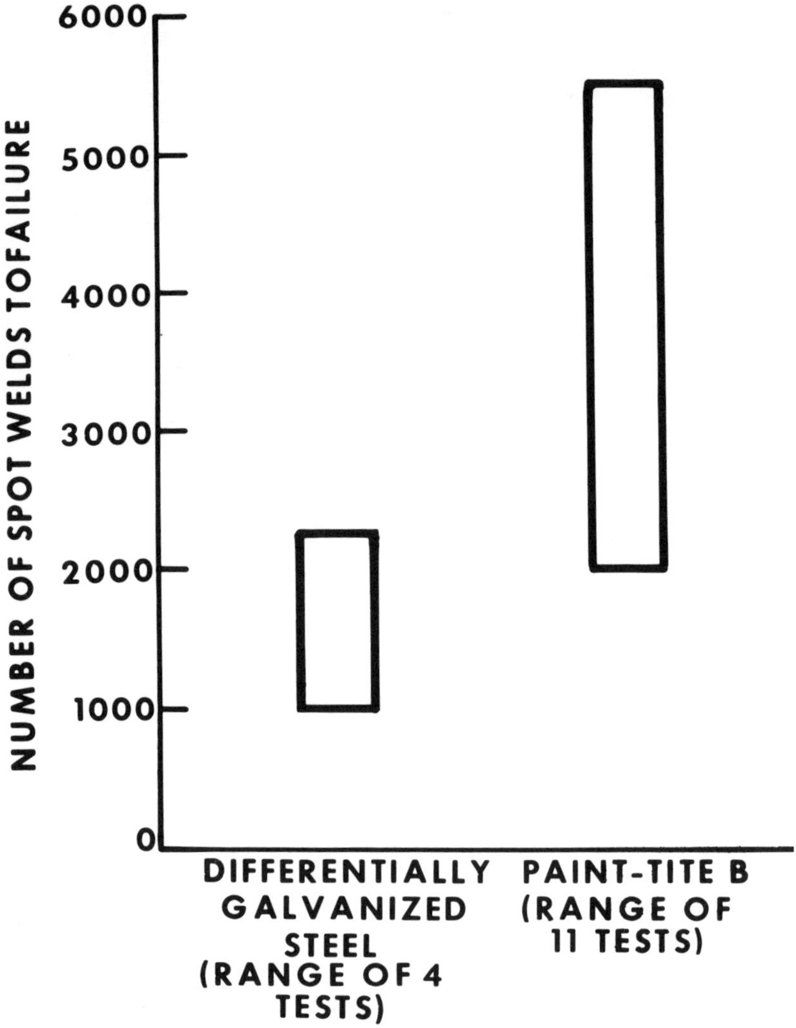

Fig. 12. Electrode Life PAINT-TITE B vs Differentially Coated Galvanized Steel

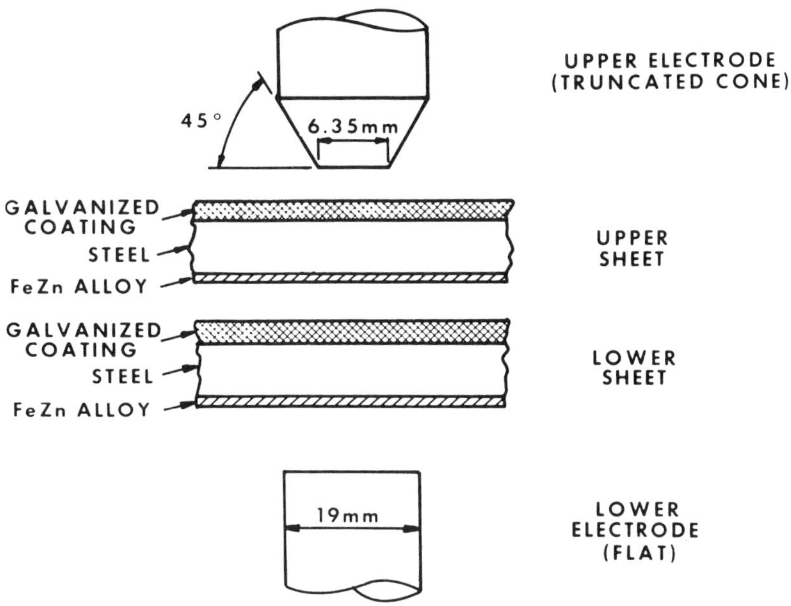

Fig. 13. Electrode Orientation for Life Test

SUMMARY

PAINT-TITE B was developed by Inland Steel with the goal of providing a suitable coated steel body panel material for satisfying the appearance, formability, corrosion resistance, painting, and welding requirements imposed by automotive companies for exposed parts. The key features of this product have been discussed, relating the production process to the properties, characteristics, and performance. The advantages of protecting the base steel on both sides, while meeting the other criteria, make this product useful for a wide range of automotive applications. To date, over 9000 Mg (10,000 T) of PAINT-TITE B have been shipped to automotive manufacturers for stamping into many exposed body panels.

CORROSION CONTROL AT ALUMINUM/STEEL JOINTS

J. S. Snodgrass
Reynolds Metals Company

INTRODUCTION

In today's efforts by the transportation industry to reduce weight through the use of a multitude of materials, joints must be made between various metals. When a less noble metal such as aluminum and a considerably more noble metal such as steel are joined, a galvanic corrosion cell or battery may be created in the joint. Techniques have been developed and proven which control corrosion in these necessary, dissimilar metal joints.

This paper will discuss in detail the corrosion problems at dissimilar metal joints. Several methods of controlling this corrosion will be presented. Finally, examples of aluminum/steel joints in railroad cars, automobiles, and trucks will be discussed.

BASICS

An understanding of the four components of the basic electrochemical corrosion cell is essential. With this understanding, the rationale for each of the control methods discussed below will be clear.

Figure 1 illustrates the four basic parts of the electrochemical corrosion cell. All four parts (1. anode, 2. cathode, 3. corrosive fluid in contact with both the anode and cathode, and 4. a metallic path connecting the anode and cathode) must be present for corrosion to occur. Complete elimination of any one of the four parts will stop electrochemical corrosion.

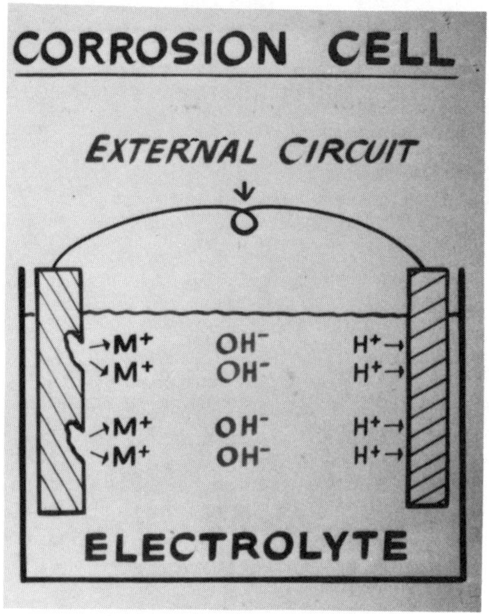

Fig. 1. Corrosion Cell

Galvanic corrosion occurs on the anode when it is a more active metal than the cathode. In the present discussion, aluminum is the more active or electronegative metal, and steel is the less active or electropositive metal.

Crevice corrosion is a localized form of corrosion that occurs in confined areas as illustrated in Figure 2. When the confined space between two pieces of metal traps and holds a corrosive fluid, such as water contaminated with salts, crevice corrosion can occur on either or both of the pieces of metal. This corrosion is driven by concentration differences of oxygen and other ionic constituents of the fluid inside and outside the crevice.

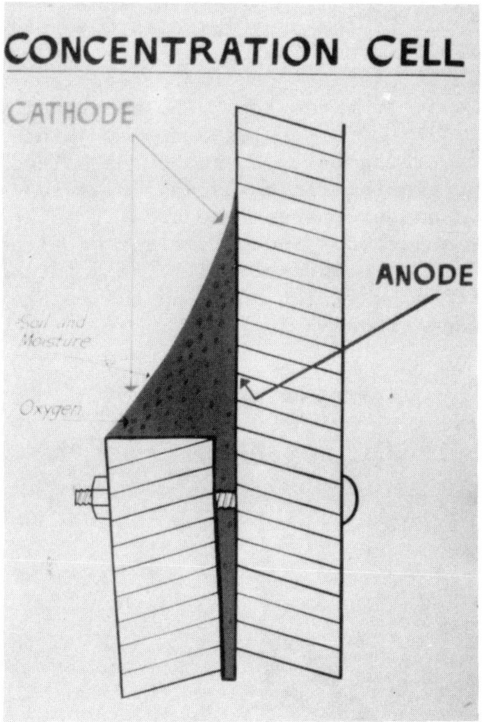

Fig. 2. Concentration Corrosion

The combination of the galvanic couple, the crevice, and a corrosive fluid can cause damaging corrosion, especially on the aluminum and occasionally on the steel. Exclusion of the corrosive fluid will stop the corrosion process.

Control of the corrosion at aluminum steel joints is accomplished by removing one of the four basic parts of the corrosion cell discussed earlier. Since the anode (aluminum) and the cathode (steel) are fixed, corrosion control must be accomplished either by breaking the electrical contact or by exclusion of the corrosive fluid.

One of the most effective and widely used corrosion control methods is the use of sealants to exclude corrosive fluids from aluminum/steel joints. Typical sealants can be divided into two categories, nonskinning and elastomeric or hardening. The nonskinning sealants, such as butyl rubbers, are used on lap joints where no movement or working will occur such as bolted and riveted joints. The elastomeric or hardening sealants, such as polysulfides and urethanes, are used in joints where some movement is expected. Enough sealant of either type should be used to completely fill the joint and permit the formation of a fillet covering the opening of the crevice. (See Figure 3.)

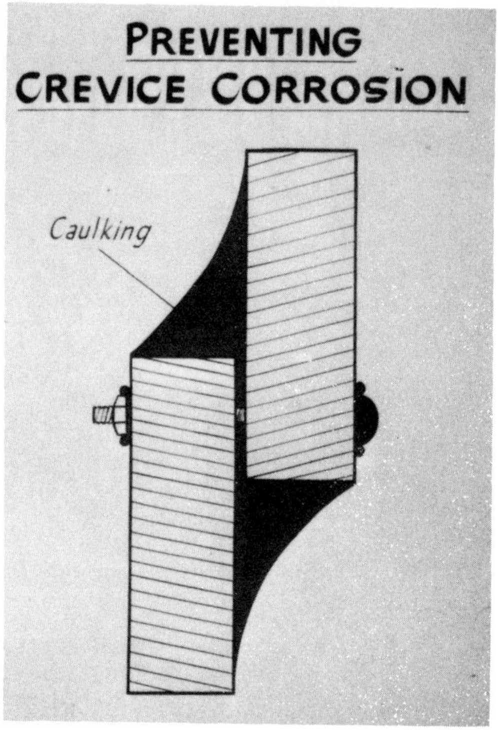

Fig. 3. Preventing Crevice Corrosion

During the design phase of assemblies which have aluminum/steel joints, consideration for corrosion prevention can pay great dividends. The location of the joint on transportation equipment can be critical. If a joint can be located away from road splash or poultice buildup, the chances for damaging corrosion to develop are greatly reduced. The configuration and orientation of the joint can influence the ability of the crevice to trap corrosive fluid. Finally, the ratio between the surface areas of the anode and the cathode can greatly influence corrosion in galvanic couples. Figure 4 illustrates what happens when a small anode is joined to a large cathode. An aluminum bolt or rivet will be severely damaged by corrosion when joined to large steel parts. On the other hand, with a steel bolt or rivet joining aluminum parts, the same quantity of aluminum will be removed as with the aluminum fastener, but it will be spread over a much larger area with minimal structural damage. Thus, galvanic and crevice corrosion problems can often be minimized by careful consideration during the design phase of a project.

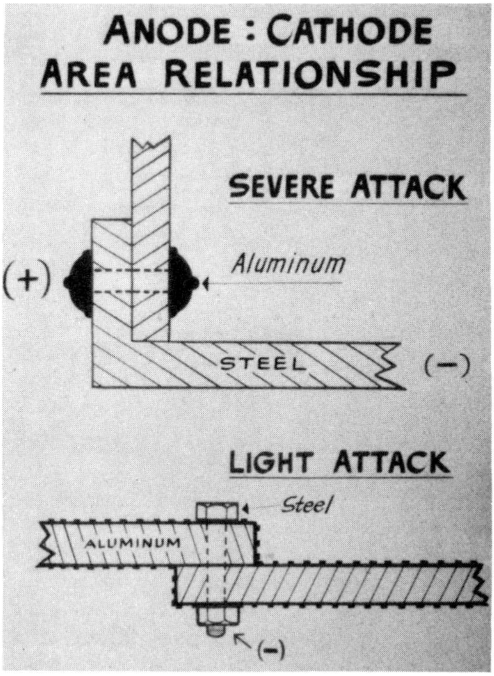

Fig. 4. Anode: Cathode Area Relationship

A favorite approach to control of galvanic corrosion is insulation or a break in the metallic path between the anode and the cathode. Figure 5 illustrates the principle of insulation in which nonmetallic material is placed between the two metals. A major problem in using this technique is locating and breaking all points of contact between the anode and the cathode. Points of contact on large parts or complex assemblies can be remote from the area of corrosion concern. Each of these points must be identified and insulated if corrosion control is to be achieved.

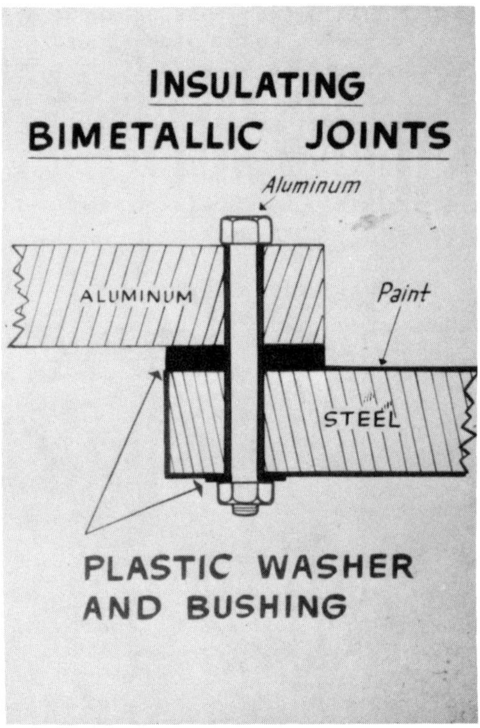

Fig. 5. Insulating Bimetallic Joints

Organic and inorganic coatings can be used to reduce the surface area of a metal which is contacted by a corrosive fluid (see Figure 5). In a galvanic couple, the surface of the cathode should be coated to achieve a more desirable area ratio. Coatings usually contain a few small imperfections or pores. Thus, coatings cannot be relied on for complete control of corrosion in galvanic couples and crevices, but they can minimize such corrosion.

EXAMPLES

Rail Cars

Many aluminum-bodied gondola and hopper cars have been built and used very successfully in coal, grain, and other services by numerous railroads for as long as twenty years. The aluminum bodies are usually attached to steel running gear. Steel and coated steel (galvanized or cadmium plated) bolt, rivet, and lock bolt fasteners have been used in these joints.

Corrosive environments of a wide variety are found in rail cars. Joints inside cars can be subject to buildups of lading residue (poultice), lading, rain, snow, and dust. Aluminum/steel joints under cars are subject to brake dust, product dust, wheel splash, and condensation.

A typical inside aluminum/steel joint is illustrated in Figure 6. The steel lock bolts hold the aluminum body to the steel center sill and the aluminum hopper gate to the steel hinges. These aluminum cars, after twelve years of coal hauling service, have experienced moderate galvanic and crevice corrosion on the aluminum adjacent to the steel. Several points should be noted. The lock bolts were originally cadmium plated and installed without any sealant, insulation, or additional coating. It is suspected that the corrosion was active only when the cars were loaded with wet coal. Because the anode (aluminum) surface area was much larger than the steel cathode, the corrosion was spread over a considerable area and had penetrated approximately 0.075" into the aluminum plate. This corrosion has caused no structural problems in twelve years. It is suggested that a sealant under the head of each lock bolt would have reduced this corrosion by filling the crevices.

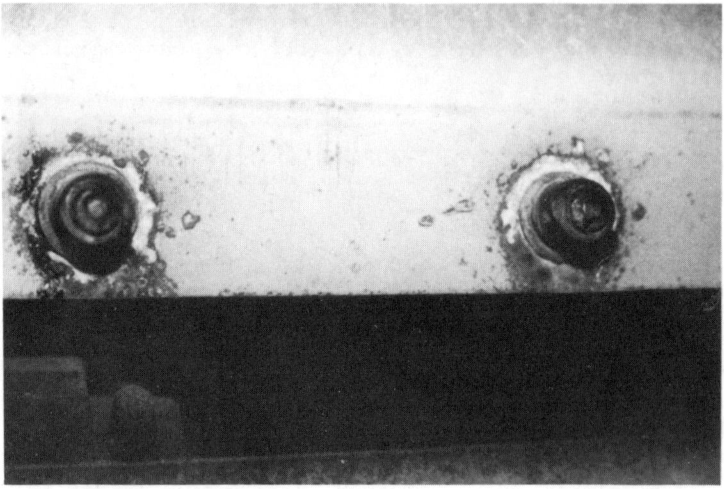

Fig. 6. Aluminum Hopper Car-Steel Bolts

The lock bolts fastening the doors to the hinges have in some cases suffered sufficient erosion corrosion to require replacement. In this case, the motion of the coal during unloading removed the cadmium plating and the corrosion products from the steel bolts. This action accelerated the normal atmospheric corrosion of the unprotected steel. The galvanic couple to aluminum was not active enough to protect the entire steel head of the fastener.

A typical outside aluminum/steel joint is illustrated in Figure 7. In the upper left corner of the figure there is an aluminum angle which is welded to the underside of the aluminum floor and riveted with steel rivets to the steel center sill. A butyl rubber sealant was used in this joint to exclude the corrosive environment. The fillet of sealant can be seen at the bottom of the aluminum angle. No corrosion was found at these joints.

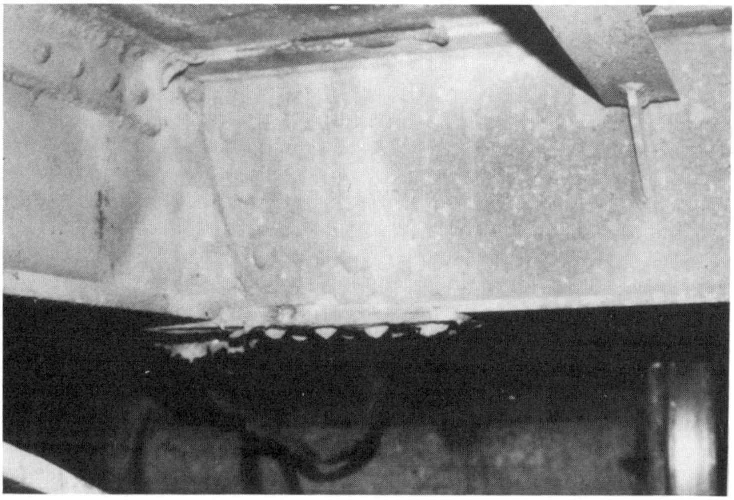

Fig. 7. Aluminum Gondola Car Body to Steel Center Sill

Automobiles

Today's automobiles utilize a tremendous number of materials which are connected and interconnected at many points. A steadily increasing number of aluminum parts are being joined to the basically steel automobile. When these joints are located in relatively protected areas such as the passenger compartment or the trunk, galvanic corrosion is not a significant concern because there is little or no corrosive fluid present.

Aluminum/steel joints on the outside of the automobiles where road splash can contact the parts and get into crevices do require corrosion control attention. Today, the aluminum/steel joints in automotive bumper systems present a real challenge to the corrosion engineer. Many combinations of aluminum and steel face bars with aluminum and steel reinforcements are attached to steel energy-absorbing units. Fasteners of several materials with various coatings are used to assemble the bumper system components and the trim and accessory items found on the bumpers. The backside of the rear bumper, especially, is subjected to severe road splash which packs salt-laden mud (poultice) into the crevices and restricted areas of the bumper assembly.

Figure 8 illustrates the mounting area of a bright anodized aluminum face bar which was bolted directly to a steel energy-absorbing (EA) unit with steel bolts. Note that there was very limited galvanic and crevice corrosion. Two major factors helped to limit the corrosion. First, this was a front bumper which was subjected to only minor poultice buildup. Second, an "insulating" tape was applied to the edges of the holes and bumper. The exposed edges did experience galvanic and crevice corrosion to the extent that the tape was undercut. The limited corrosion control resulting from the use of the insulating tape is a typical shortcoming of this approach to controlling galvanic and crevice corrosion at aluminum/steel joints. A butyl rubber sealant in this joint could have excluded the corrosive fluid and minimized the corrosion. In mass production, there is often a reluctance to use sealants because proper application is required to get maximum corrosion protection.

Fig. 8. Bright Anodized Aluminum Bumper at Joint With Steel

Fastener selection for aluminum bumper systems is an area of concern. Fasteners invariably are steel and have many crevices associated with their function. Figure 9 illustrates an extremely severe example from a testing program of a chromium-plated aluminum bumper with a steel fastener. The following events probably contributed to this corrosion problem. The bolt was tightened to a point that the chromium plating system was damaged around the head of the bolt. The assembly was exposed to an accelerated test environment which included continuous wetness with salt present. Thus, galvanic couples became active between the steel bolt and the aluminum and between the copper of the plating system and the aluminum. The result was severe galvanic and crevice corrosion.

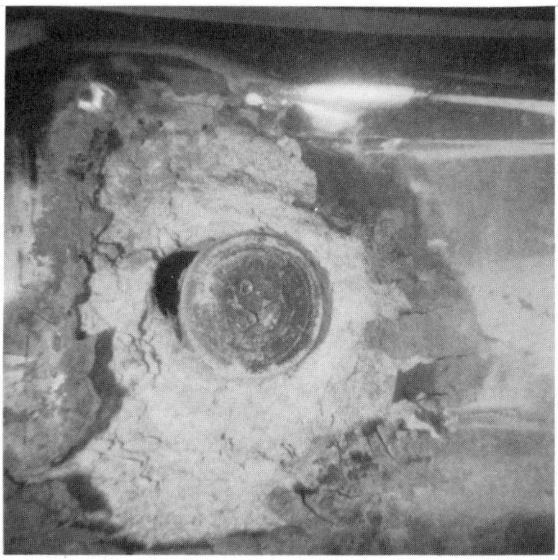

Fig. 9. Steel Bolt in Chromium Plated Bumper

This corrosion could have been controlled by the use of a sealant to keep the corrosive fluid out of the joint or by the use of a soft, nonwicking gasket under the head of the bolt to prevent damage to the plating and ingress of corrosive fluid. An often-tried approach is coating of the steel fasteners. Unfortunately, organic coatings are often broken during assembly, especially in the threads where crevices exist. Metallic coatings, such as zinc (galvanizing), provide limited protection. Once the metallic coating is consumed by corrosion, one is back to bare steel. Yet another approach is the use of more expensive alternate materials for fasteners such as aluminum or stainless steel.

There is no clearly best approach to this fastener problem. By keeping in mind the alternatives discussed in the preceding paragraph and the basic corrosion control techniques discussed earlier, one can achieve reasonable corrosion control.

Trucks

Aluminum side rails on highway trailers are often connected to steel cross members. Figure 10 illustrates an example of the galvanic and crevice corrosion that can occur, especially around the wheels. The wheel area is subject to frequent splashing with water, deicing salt, and mud. The buildup of a poultice around the aluminum steel joints is inevitable.

Fig. 10. Aluminum Trailer Rail Corrosion

Several alternatives can be considered to control this problem. Coatings and sealants can be liberally used in the areas subject to splash. A baffle or splash shield can be installed to keep the splash from contacting the structural joints. Because this problem is confined to a relatively small part of a trailer, the suggested additional corrosion control measures need be applied only where experience indicates a need.

A final example will be used to illustrate how laboratory testing can be used to help select the most promising candidate aluminum alloy for use in a galvanic couple. The project was the development of a prototype aluminum pickup truck frame. The galvanic couples of concern were the attachments of steel suspension, drive train, and body parts to the aluminum frame. This example also illustrates the fact that aluminum alloy selection can play a part in the control of corrosion in aluminum/steel joints. It is not, however, the purpose of this paper to discuss the corrosion resistance characteristics of each of the aluminum alloys used in transportation equipment.

In this test program, coupons of two aluminum alloys were fastened with steel rivets to steel coupon racks. The coupon configuration illustrated in Figure 11 created crevices and galvanic couples. The steel rack was painted in all cases using a conventional truck frame coating system. In addition, approximately half of the coupons were painted after assembly.

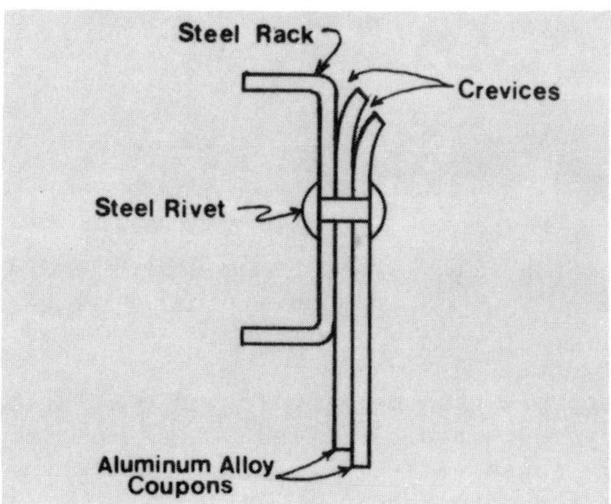

Fig. 11. Truck Frame Corrosion Test Coupon Arrangement

Figure 12 illustrates the condition of typical coupons of the two aluminum alloys after exposure in a 90-day alternate immersion test. This test exposed the coupon racks and coupons to alternate wet and dry conditions. During 10 minutes of each hour the coupon racks were immersed in a 3.5% NaCl solution. During the remaining 50 minutes of each hour the coupon racks were hanging in air at a temperature of 80°F and 50% relative humidity. This is a severe test. Note aluminum alloy B was relatively unaffected by the corrosive environment and the galvanic couples. The machined areas on alloy B were the results of efforts to remove the steel rivet. This machining did not destroy any significantly corroded areas. Thus, aluminum alloy B appears to be a satisfactory choice for a pickup truck frame with steel attachments.

Alloy A Alloy B

Fig. 12. Truck Frame Corrosion Test Coupon Typical Results

Prototypes have been built and initial test track durability test results indicate that the laboratory test was a very conservative predictor. No corrosion problems have been found to date.

CONCLUSIONS

Three key points have been emphasized in this paper.

1. Aluminum will be joined to steel in transportation equipment.

2. Corrosion can occur at aluminum/steel joints when they are neglected.

3. Corrosion can be controlled at aluminum/steel joints through the use of one or more of the following: (a) sealants; (b) design; (c) insulation; (d) coatings; and (e) materials selection.